"*When Brains Dream* unveils a novel neuroscientific model comprising an elegant and surprising piece of the puzzle of why we dream the way we do. If you are curious about the curiouser and curiouser qualia of dreams, read this book!"   —Stephen LaBerge, author of *Lucid Dreaming*

"Bob and Tony have an answer to a problem that's been puzzling people for 200 years: why do we dream? They explain why we daydream too; why it's vital for sanity, so for me *When Brains Dream* has acted as an intervention."   —Mary Wakefield, *Spectator* (UK)

"The book wields dreamy anecdotes and complex neuroscience to try to grasp the importance of these phantasms [dreams]."
          —*Scientific American*

"Every night, we turn out the light and go to the movies. For as long as humans have been conscious of our world, we've wondered what's going on in that other world in which we spend a third of our lives. Some of my strangest dreams have been while I was asleep in Bob Stickgold's lab. Finally, in *When Brains Dream*, I have a way to understand them. There's more here than you ever dreamed of."
          —Alan Alda, *New York Times* best-selling author of
          If I Understood You, Would I Have This Look on My Face?

"*When Brains Dream* steers a reasonable and broad-minded course between the many interpretive whirlpools that have swallowed previous explorers of dreams."   —Nick Romeo, *Washington Post*

"Where do we go in our dreams at night? And why do we dream in the first place? Are we the only species that dreams? *When Brains Dream* provides a truly comprehensive, scientifically rigorous and utterly fascinating account of when, how, and why we dream."
          —Matthew Walker, author of *Why We Sleep*

WHEN BRAINS DREAM

*Understanding the Science
and Mystery of Our
Dreaming Minds*

ANTONIO ZADRA

AND

ROBERT STICKGOLD

W. W. NORTON & COMPANY
Independent Publishers Since 1923

For information about permission to reproduce selections from this book, write to
Permissions, W. W. Norton & Company, Inc., 500 Fifth Avenue, New York, NY 10110

For information about special discounts for bulk purchases, please contact
W. W. Norton Special Sales at specialsales@wwnorton.com or 800-233-4830

Manufacturing by LSC Communicatoins, Harrisonburg
Book design by Lovedog Studio
Production manager: Lauren Abbate

The Library of Congress has catalogued the hardcover edition of this book as follows:

Names: Zadra, Antonio, author. | Stickgold, R. (Robert), author.
Title: When brains dream : exploring the science and mystery of sleep /
Antonio Zadra and Robert Stickgold.
Description: First edition. | New York, NY : W. W.Norton & Company, [2021] |
Includes bibliographical references and index.
Identifiers: LCCN 2020027749 | ISBN 9781324002833 (hardcover) |
ISBN 9781324002840 (epub)
Subjects: LCSH: Sleep—Research. | Sleep disorders.
Classification: LCC RA786 .Z33 2021 | DDC 613.7/94072—dc23
LC record available at https://lccn.loc.gov/2020027749

ISBN 978-1-324-02029-5 pbk.

W. W. Norton & Company, Inc., 500 Fifth Avenue, New York, N.Y. 10110
www.wwnorton.com

W. W. Norton & Company Ltd., 15 Carlisle Street, London W1D 3BS

1 2 3 4 5 6 7 8 9 0

*This book is dedicated to those early
dream explorers whose innovative methods
and remarkable insights laid the foundation for
the scientific study of dreams and to all those who have
ever wondered why we dream, where dreams come
from, and what meanings they may hold.*

CONTENTS

PREFACE

....

WHAT ARE DREAMS? WHERE DO THEY COME FROM? What do they mean? And what are they for? Humanity has been trying to answer these questions for thousands of years, without much success. But since the nineteenth century, scientists have been asking these questions anew, attempting to unravel the relationship between brains, minds, and dreams. Now, in the twenty-first century, we may be close to an answer.

Like everyone else, you will come to this book with your own preconceived notions about dreaming. To some people, the idea of "a science of dreaming" feels like an oxymoron or even an outright impossibility. Scientists study things and processes that we can see and measure. We study the observable, quantifiable world we live in, from the infinitely small out to the edges of the universe. In contrast, dreams are subjective events, invisible to all but the dreamer, unknowable to others except by the fragmented and often fuzzy recollections that a dreamer can tell us. For other people, there is a mystery and wonder inherent in our dreams that scientific explanations can only diminish or even destroy. Some believe that science has already shown that dreams are merely the meaningless reflections of the random firing of neurons in the sleeping brain. Nothing, we believe, could be further from the truth, and we argue almost the exact opposite of each of these claims.

We began studying dreaming in the early 1990s and, between the two of us, have published over two hundred scientific papers on sleep and dreams. But for both of us, the mystery and wonder of dreaming continues to grow. In fact, our long-standing fascination with the universal *experience* of dreaming—why and how people dream—is at the heart of our decision to write this book. The wealth of recent discoveries and insights about the sleeping brain and the nature of dreams suggests that dreams are psychologically (and neurologically) important and meaningful experiences.

In this book we start by examining how, as children, we only gradually come to understand what dreams and dreaming are. We then take you on a tour of dream research, describing the contributions of early dream explorers from the nineteenth century, whose pioneering methods and ideas foreshadowed many scientific approaches to the study of dreams. We look at the works of Sigmund Freud and Carl Jung, presenting them in ways that may be unfamiliar to you. You'll learn about the discovery of REM sleep (when our most vivid dreaming occurs), our current understanding of the functions of sleep, and debates surrounding the possible functions of dreams. We will tell you what we know about who dreams, when they dream, and what they usually dream about, and we'll explore the question of whether other animals dream. We also delve into the content of typical dreams, recurrent dreams, sexual dreams, and nightmares. Then we'll examine how dreams facilitate creativity, learn how they can be used for personal insight, and explore the world of lucid dreaming as well as claims of telepathic and precognitive dreams. And perhaps most exciting, we'll offer new insights into why we dream.

Pulling together a variety of compelling neuroscientific ideas and state-of-the-art findings in the fields of sleep and dream research, we propose a new and innovative model of why we dream. We call this model NEXTUP, which stands for "Network Exploration to Understand Possibilities." By detailing the workings of NEXTUP, we show why the human brain needs to dream, and we offer new answers to all four of our original questions—what dreams are, where they come from, what they mean, and what they're for.

Along the way, we'll provide evidence regarding a number of claims about dreaming: We don't only dream in REM sleep, nor do we all dream in black and white. Additionally, our dreams are rarely driven by repressed desires. Dreams can, in a real sense, predict the future in a way that we can't when we're awake. Furthermore, dreaming has a cognitive basis; there are reasons for why our dreams feel so real and meaningful. You'll learn what blind people dream about and where the brain gets the images and concepts that it uses to create dreams. We'll provide what we think are novel insights into nightmares and a host of other dream-related disorders. You'll be better able to understand your dreams as well as those of others. And you'll learn that even with all these advances in our knowledge, ample mysteries and magic still surround the world of dreams to keep us all in wonder.

We hope you enjoy reading this book as much as we enjoyed writing it.

WHEN BRAINS DREAM

Chapter 1

THINKING
ABOUT DREAMS

"DREAMS: *NOUN, PLURAL*: A SERIES OF THOUGHTS, images, or emotions occurring during sleep." This is how most dictionaries define the topic at the heart of this book. While this definition gives us a reasonable starting point to explore questions about what dreams are and how we might think about them, it raises more questions than it answers. For instance, where do these thoughts, images, and emotions come from? How do they relate to the ones we have while we're awake? And why do we dream at all?

These are the kinds of questions we'll be addressing in this book, but they are by no means original. Questions about the origins and meaning of dreams are as old as humankind. From the 4,000-year-old *Epic of Gilgamesh*—the world's oldest recorded story—to ancient Greek philosophy and the birth of modern medicine, dreams have always held a privileged, if misunderstood, place in human history. Dreams figure prominently in many ancient texts, including the Old Testament, the Talmud, the Upanishads (Indian spiritual treatises written between 1000 and 800 BCE), the Bardo Thodol (also known as the *Tibetan Book of the Dead*), Homer's *Odyssey* and *Iliad*, Hippocrates' *On Dreams*, and no fewer than three works of Aristotle—*On*

Dreams, On Sleep and Waking, and *On Prophesying by Dreams.* Like many other classic texts, these monumental works have had much to say about dreams, and in most cases they offered definitive—but often contradictory—answers to compelling questions. Are dreams omens? Are they communications from the gods? Do they represent a truer and higher reality? And how do we understand their meaning? The answers to these questions put forth by our ancestors did more than influence people's beliefs about dreams; they profoundly affected how people perceived the human condition and our role in a largely unknown universe.

It is not surprising, then, that numerous scholarly reviews[1] of the historical literature on dreams—from records of the earliest known civilizations to twentieth-century writings—have found that dreams played a key role in the establishment of all major religions, in our conceptualization of the cosmos, the nature of death, and the ways in which the secular world can intersect that of the sacred and divine. In a very real sense, dreams have shaped our conception of the world, including our place within it.

At a more personal level, consider the idea that dreams almost certainly have influenced how *you* came to understand the world around you. Don't believe us? Let's examine how we all gradually (and not without difficulty) came to understand what we as adults mean by dreams and dreaming.

Most of us likely first heard the word *dream* associated with a negative statement: "No, honey, it was just a dream!" But for our younger selves, hearing that wasn't very helpful. A monster, a dinosaur, a bad man, had been right there in front of us, and our parents tell us it was just a dream? What in the world is a dream? What did they mean, "just a dream"? Just a toy? Just a ghost? Just someone dressed up like a monster? Just something scary (but hopefully harmless) that happened to live in our closet or under our bed? What could that word possibly mean? By now we are all so famil-

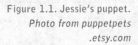

Figure 1.1. Jessie's puppet.
Photo from puppetpets
.etsy.com

iar with the idea of dreaming that we forget what a strange and unlikely concept it is.

We can think about dreams in many different ways, each of which makes a certain sense. Perhaps the earliest and simplest explanation was that what we experience in our dreams really happens. When Bob's daughter Jessie was two years old, her grandfather Irv gave her a wood-and-yarn puppet of a duck. He made it walk around, say "Quack, quack," and give her a kiss. Jessie loved it, and she asked Bob and Irv to hang the puppet on the wall in her bedroom before she went to bed.

But later that evening, Bob heard screams of terror coming from upstairs, and flew up to Jessie's room to find her standing up, pressed against the crib railing, her arms stretched toward him in panic. As Bob picked her up, she spun around in his arms, looked back down into her crib, and screamed, "There's a duck in my bed!"

We adults know that dreams can seem that real. At Jessie's age, it takes a while to learn that they're not actually real. Indeed, the road to an adult understanding of what dreams mean is anything but simple.

When Tony's nephew Sebastian was five—more than twice as old as Jessie when she got her duck puppet—he had a pretty good grasp of what dreaming was all about. But Sebastian's understanding couldn't stand up to Tony's questioning. Here's how Tony remembers the discussion.

T: Tell me, Sebastian, do you sometimes remember your dreams?

S: Yes.

T: Good. And where do these dreams take place?

S: Hm. Well, they're like in front of me.

T: In front of you how?

S: They're just in front of me. I see them with my eyes.

T: And when you have these dreams, are you sleeping or awake?

S: (Gives Tony a strange look.) Sleeping.

T: Okay. And do you sleep with your eyes open or closed?

S: Closed, of course.

T: So if your eyes are closed during your sleep, how do you manage to see your dreams?

S: (A pause) Mom! Uncle Tony is asking me those stupid questions again!

Poor Sebastian. Tony kept it up for several more years, but we can learn a few lessons from this. The first is that having a dream researcher in the family isn't always fun. The second is that although Sebastian was five years old, he was confused when he realized that he could see images even when his eyes were closed, and he didn't know how that could be. Think about Tony's question. How would you answer it? Probably you'd explain that it's like when you close your eyes and imagine something, and you'd be right—kind of. But that doesn't answer the question. How can you even do it when you imagine something? And why is it ten times more vivid and believable when you dream it than when you imagine it? When the brain dreams, how does it create an *experience* of seeing, hearing, or feeling something that's as realistic as in waking life? Sebastian is not alone in being con-fused, and to do justice to the story of dreaming, we must address this problem of what dreams *are*.

To understand the concept of dreaming, children (as well as adults) require more insights about how the world works than initially meets the eye—including the mind's eye. And, whether you can remember it or not, you probably went through a very similar process as you tried to figure out what these things called "dreams" actually were.

As part of his pioneering work on children's cognitive develop-ment, the great Swiss psychologist Jean Piaget systematically inves-tigated children's understanding of dreams as they grew up. What he found was that most preschoolers believed that dreams were real, originated from outside the dreamer, and could be seen by others. It wasn't until sometime between the ages of six and eight that most children grasped the idea that dreams were not only imaginary, but they could not be observed by others. And it was only around the age of eleven, according to Piaget, that children came to fully understand the nonphysical, private, and internal nature of dreams.

Some thirty years later, in 1962, Montreal-based researchers Monique Laurendeau and Adrien Pinard carried out one of the most extensive and widely cited studies[2] of children's conceptions of dreams. As part of their larger investigation into children's development of causal thinking, Laurendeau and Pinard asked some five hundred children, aged four to twelve, various questions about dreams, includ-ing, "Do you know what a dream is?" and "While you are dreaming, where does your dream take place?" In line with Piaget's observations, these researchers described four stages of comprehension, designated Stages 0 to 3. In the first stage, which characterized about half of the four-year-olds, the child does not comprehend what a dream is, nor even understand the question. Later, in the next stage, the child believes that dreams are as real as waking-life experiences, that dreams exist separately from the dreamer, and that an observer in her room would be able to witness her dreams. The third stage, reached by less than half of the six-year-olds, is an intermediary step during which the child goes from a view of the dream as an external event happening "in front of my eyes" or "in my bedroom" to the idea that dreams are like movies that take place "in my head." Finally, in the fourth stage, the child (usually by ages eight to ten) fully understands that dreams are internal, private, and imaginary mental experiences.

More recent work suggests that some children understand that dreams are not part of the external world as early as ages three to five (we'll examine *when* children start to dream and *what* they dream in Chapter 6). Even so, the steps involved in attaining an adult-like conception of dreams remain the same: we first believe that dreams are part of the real world, then grasp their unreality, then their private nature, and finally locate the dream as taking place within our mind. Some variations exist. For example, some children will say that dreams originate from "inside my head" but still think their dreams are real and visible to outside observers, such as people sleeping nearby or others present in their bedroom. Other children understand the private and subjective nature of dreams, but they still maintain that the dream world is physically real—at least while they are in it. Of course, some adults believe this, too.

Moreover, even after understanding the unreal, private, and internal qualities of dreams, some children, when asked where dreams come from, answer that dreams originate from the air, the sky, or the night. Depending on their culture and beliefs, some children will say that dreams come from supernatural sources, such as God or heaven. As Laurendeau and Pinard pointed out, "Far from being characteristic of the most primitive beliefs, the recourse to divine or supernatural being may be observed at all levels. Indeed, even in children who hold the strongest convictions about the subjectiveness and individuality of dreams, a reference to the action of a divine power may often occur."[3]

In many ways, the development of children's conceptions of dreams mimics how society's conceptualization of dreaming has evolved across millennia. Perhaps, as we transition out of early childhood, we come to believe that we understand the nature and origins of dreams when, in fact, we've merely internalized what our elders have told us about what dreams are and where they come from. What if instead of learning that dreams take place "inside your head," we had been told that dreams are sent to us by higher powers to guide or deceive us? Or that dreams take us to real places in physical or spiritual worlds?

Cognitive milestones are clearly needed to achieve an adult understanding of the modern-day concept of dreaming. But our own dream experiences, coupled with what our parents, friends, and society tell us about dreams, all play a role in how we ultimately come to view these perplexing nocturnal experiences, including their origins, worth, and meaning—or lack thereof.

Let's return for a moment to the idea that an understanding of dreams requires the ability to distinguish between real and imagined events. As adults, we sometimes recall a memory and ask ourselves, "Did that really happen, or did I dream it?" Usually these are vague recollections, hard to place in space and time. But sometimes, and for some people, they're very clear. A colleague of Bob's, Tom Scammell, is a physician and researcher who studies the sleep disorder narcolepsy, a neurological disorder that affects the control of sleep-wake cycles. He came to Bob one day and described a narcolepsy patient, a young girl, who had thrown herself down a flight of stairs to show her brother how she could fly. As you might expect, she fell heavily down the stairs. But she was not to be dissuaded; she was so sure she could fly, she climbed back up the stairs and tried again! She had had a dream of flying that was so real, she was sure it had actually occurred. "I think that happens a lot," Tom told Bob.

It turns out he was right. Together with Tom and other colleagues in Boston and the Netherlands, Bob conducted a study[4] that put this question to 46 narcoleptic patients and another 41 "control" participants: "Have you ever had the experience of being unsure whether something was real, or if it was from a dream?" In follow-up interviews, the researchers made clear that they were talking about confusions that lasted for at least hours, and for which patients sought out additional information to clarify whether the event had actually happened. In the end, only about 15 percent of the control participants described ever having been unsure whether something was real or from a dream, and only two people reported having this experience more than once in their lives. In contrast, over three-fourths of the patients described such instances, and all but one of them reported it happening at least once a month. In fact, two-thirds said it happened at least once a week!

Although narcolepsy patients generally report more vivid dreaming than other individuals, those patients who reported these experiences did not report more vivid dreaming than those who did not, so the confusion cannot be attributed to more vivid dreaming. Instead, the problem might result from abnormally strong memories of the dreams being created because of neurological and neurochemical abnormalities associated with the disorder. As of this writing, we still don't know why patients with narcolepsy suffer from this malady.

But we do know that the confusion is intense and has major effects on their lives. One of the patients in the study, after dreaming that a young girl had drowned in a nearby lake, asked his wife to turn on the local news in full expectation that the event would be covered. Another patient experienced sexual dreams of being unfaithful to her husband. She believed this had actually happened and felt guilty about it until she chanced to run into the "lover" from her dreams and realized they had not seen each other in years and had never been romantically involved. Several patients dreamed that their parents, children, or pets had died, believing that this was true (one patient even made a phone call about funeral arrangements) until, to their great shock and relief, the presumably dead person suddenly reappeared.

Confusion between dreams and wakefulness isn't limited to people with narcolepsy. False awakenings, in which people *dream* that they have woken up (often in their habitual sleep environment) can happen to anyone. In these dreams within dreams, people may "wake up," get out of bed, shower, and be about to fix themselves breakfast when, to their utter astonishment, they wake up again! People who've experienced a false awakening are often amazed by the dream's exquisitely rendered details that helped fool them into believing they were awake. The dream *looked* and *felt* incredibly real to them. So real, in fact, that they mistook it for reality.

In his home-based studies on dreams, Tony has collected over 15,000 dream reports from hundreds of men and women. In several of these studies, people have spontaneously noted waking up from a dream in the middle of the night, taking the time to write down the recalled dream before going back to sleep, and then finding their dream diary blank in the morning! In these cases, people *dreamed* that they woke up and recorded their previous dream (either in writing or via voice recording), and then *woke up* convinced that what had in fact been a dream had actually occurred.

The uncertainty surrounding the states of dreaming and wakefulness was perhaps best captured by the influential Chinese philosopher Zhuangzi (369 BCE to 286 BCE). In his famous *Butterfly Dream*, he wrote: "Once upon a time, I, Zhuangzi, dreamt I was a butterfly, fluttering hither and thither, to all intents and purposes a butterfly. I was conscious only of my happiness as a butterfly, unaware that I was Zhuangzi. Soon I awakened, and there I was, veritably myself again. Now I do not know whether I was then a man dreaming I was a butterfly, or whether I am now a butterfly, dreaming I am a man."

Our conclusions from all this information are that when brains dream, the dreams they create are not only believable while we're dreaming them but also can be equally believable afterward. Perhaps it's not surprising then that, depending on individual circumstances, we can think of dreams as either real life (what Jessie believed about the duck and the confusion of narcoleptics); portals into equally real or alternate worlds; messages and prophecies from the gods; unfulfilled wishes; random brain noise; nocturnal entertainment; communications from the future, the dead, or other minds; sources of personal insights, problem solving, and creativity; or a window into memory processing.

When brains dream, they leave us with all of these possible explanations for what is happening. In this book, we will discuss them all, and see where these alternatives take us. We'll find that there's not just one right answer, and that they aren't necessarily mutually exclusive. With little stretches (or sometimes big ones), all of them can be reasonable. But for us, the most exciting explanation, both intellectually and scientifically, is the last one about memory processing. We'll have much more to say about it later on, but here's a preview. We know now, based on almost twenty years of research, that while we sleep, the brain is constantly working, processing our memories from the day just past. For every two hours we spend awake, taking in new information, it appears that the brain needs to shut down all external inputs for an hour to make time to figure out *what it all means*.

Bob's first computer was an Apple II+. It had 48 kilobytes of memory. That's right—kilobytes. That's 0.048 megabytes, or 0.000048 gigabytes, or 0.0001 percent of the memory in his iPhone. And its CPU ran 2,400 times slower. Despite its limitations, that computer could memorize everything Bob typed on its keyboard, music that came in from a cassette recorder, or pictures drawn on a primitive tablet. It just couldn't begin to tell him what it meant. It probably couldn't even hold the concept of meaning. It is only in the last few years, with 10-terabyte hard drives and the introduction of new artificial intelligence (AI) and "deep learning" programming techniques, that computers have begun to answer the question of what the information they collect "means." That's the hard part of the job, whether you're a computer or a human, and our brain does the hardest part of it while we sleep. The calculations our brains perform during sleep are almost unbelievable. As for dreaming, we now believe that our brains slipped a bit of consciousness into the mix, in the form of dreaming, to help with this amazing process. We'll describe how dreaming does this in Chapters 7 and 8.

Before closing this chapter, let's return to one of the core aspects of dreaming that young children often struggle with. After figuring out that some things in life are real and others unreal, and distinguishing things that exist in the physical world from those whose nature is immaterial, children need to grasp that while *they* can observe their dreams, the images they see cannot be witnessed by anyone else. The private nature of dreaming has vital implications for anyone interested in other people's dreams. It doesn't matter if you're a neuroscientist, clinician, pastor, or concerned parent, you can never study someone else's dream experiences directly. All you have access to is the *description* of the experience the other person provides, whether it is shared verbally, in writing, through a drawing, or in a piece of performance art. Thus, the concept of dreams refers not only to "the series of thoughts, perceptions, or emotions that are experienced during sleep" but also to what people *remember* of these experiences and the spoken or written *reports* the dreamer eventually provides based on her (often short-lived) memory of the dream.

The upshot of all this is that both as a concept and an actual experience, dreams are much trickier than most people imagine. And, to make things worse, there is no consensus among researchers as to what even counts as a dream. Interdisciplinary groups from the International Association for the Study of Dreams and the American Academy of Sleep Medicine concluded that "a single definition for dreaming is most likely impossible given the wide spectrum of fields engaged in the study of dreaming, and the diversity in currently applied definitions."[5] Thus, depending on one's perspective, dreaming can be synonymous with the term *sleep mentation*, which refers to the experience of *any mental activity* (perceptions, bodily feelings, isolated thoughts) during sleep, or it can be restricted to more elaborate, vivid, and story-like experiences recalled upon awakening.

In this book, we'll adopt a broad view of dreaming that encompasses everything from fleeting, fragmented, and thought-like forms of sleep mentation all the way to dramatic, seemingly epical nocturnal adventures. In general, however, our focus will be on the more complex and involving forms of dreaming—those rich, immersive experiences that have given dreams their sense of mystique and have intrigued and perplexed humans from time immemorial.

Chapter 2

GRASPING AT DREAMS

EARLY EXPLORERS OF THE DREAM WORLD

WHEN WE ASK STUDENTS, FRIENDS, OR AUDIENCES AT our public lectures what marked the birth of the scientific study of dreams, most of them answer that it was Freud. (In rare cases, people answer that it was the discovery of REM sleep.) Indeed, Freud's influence on our beliefs about dreaming cannot be overemphasized. Consider this quote:

> *The perception or imagination of a thing may rouse a momentary desire which we repress as foolish or wrong. The next night these half-formed psychical tendencies, relieved of all their restraint, work themselves out. . . . We may assume, perhaps, that in each case the dream was the expansion and complete development of a vague fugitive wish of the waking mind . . . the dream becomes a revelation. It strips the ego of its artificial wrappings and exposes it in its native nudity. It brings up from the dim depths of our subconscious life the primal, instinctive impulses. . . . Like some letter in cipher, the dream inscription when scrutinized closely loses its first look of balderdash and takes on the aspect of a serious, intelligible message.*

Who could better sum up the core of Freud's theory of dreaming? Apparently the English psychologist James Sully could, because he wrote and published this quote seven years before the appearance of Freud's *The Interpretation of Dreams*. Sully had much to say about the origins and interpretation of dreams, as reflected in his 1893 article, "The Dream as Revelation."[1] His writings contained many of the same elements that Freud would go on to use in his own model of dreams—something Freud would not acknowledge until more than twenty years later in the fourth (1914) edition of *The Interpretation of Dreams*.

In fact, the decades preceding Freud's *The Interpretation of Dreams* saw a number of researchers conducting a variety of innovative investigations into the nature of dreams. Moreover, many of the modern neuroscientific ideas about sleep and dreams presented throughout this book originated not in the works of Freud or Jung, but in studies conducted by even earlier explorers of the dream world, explorers whose names and work have been largely neglected and forgotten. To this day, most introductory and even advanced texts on the science of dreams begin with one author: Freud. Rarely do they mention the extensive and wonderful work on dreams carried out in the late nineteenth century. And it's obvious why.

In his opening chapter of *The Interpretation of Dreams*, Freud presented a highly influential review of the pre-twentieth-century scientific literature on dreams. For decades to come, this chapter summarizing the work of fifty other authors remained *the* definitive resource for anyone interested in the pre-Freudian history of dream research. Freud's literature review certainly helped bring the long-standing scientific interest in dreams to people's attention, but careful examination by a host of researchers and historians[2] of Freud's writings *and* of the works he cited has brought several key observations to light.

First, many of Freud's ideas about dreams were not as original as he made them out to be; several, in fact, were based on the work of others whose contributions he failed to properly acknowledge. For instance, as evidenced by Sully's quote above, many authors before Freud had suggested that dreams sometimes reflect wishes, including repressed ones; likewise, others had already surmised some of the mechanisms he invoked to explain how dream images were formed.

Moreover, Freud was unjustified in his strongly dismissive assessment of the research on dreams conducted before publication of *The Interpretation of Dreams*. Specifically, he exaggerated the inconsistencies between the views of different authors, minimized the importance that "medically oriented" researchers had given to psychological factors implicated in dreaming, and misrepresented many researchers' views by claiming they had been interested only in physical sources of dreams, or in what at that time were known as somatic theories of dreams.

In a claim that was misleading at best, Freud also presented himself as the "founder" of the psychological (as opposed to medical) study of dreams. As we're about to see, several authors before him had put forth many innovative ideas regarding the psychological dimensions of dreams.

Finally, Freud himself admitted to his aversion to reading other people's work on dreams, going as far as to write, "The literature [on dreams] which I am now reading makes me completely stupid. A horrible punishment for those who write."[3]

For all these reasons, Freud's opening claim that his examination of the literature preceding *The Interpretation of Dreams* had led him to conclude that "the scientific understanding of dreams has made very little advance"[4] constitutes an unfair, biased, and self-serving assessment of this body of work.

G. W. Pigman, a professor at the California Institute of Technology known for his work on the history of psychoanalysis, may have said it best in his detailed analysis of Freud's literature review. He wrote that this chapter "makes Freud's own theory appear more revolutionary than it actually is. Freud exaggerates the dominance and neglects the complexity of physiological theories of dreams; he also underemphasizes the tradition of the dream as revelation.... Freud was not, as he claimed, the only scientist or physician of his day to believe that dreams are interpretable and meaningful."[5]

By characterizing much of the work conducted before him as relatively trivial and purely physiological or medical in nature, Freud was better able to promote his claim of the trailblazing nature of his own psychological theory of dreams. In time, dreams became almost the exclusive purview of Freudian psychoanalysts, and people grew more interested in questions about how a specific dream could be interpreted than in how the sources and content of dreams could be scientifically investigated. Thus, thanks to Freud's dismissive view of their importance, studies that focused on things other than dreams' "hidden" meanings were either forgotten or ignored as being of little interest.

To be clear, and as we'll see in the next chapter, Freud's ideas and theories were revolutionary and the impact of *The Interpretation of Dreams* immeasurable. But his contemptuous and selective portrayal of the dream-related ideas and studies that preceded his magnum opus, combined with the importance and myths that came to surround his work and the psychoanalytic movement itself, obscured the valuable contributions of dozens of others before him and effectively quashed most scientific dream research for the next fifty years. With time, pre-Freudian research on dreams was relegated to the dustbin and eventually forgotten by most. So, we will begin our journey into the science of dreams by returning to some of these neglected pioneers of dream research. Let's give credit where credit is due.

For centuries, dreams were explained within religious and metaphysical belief systems and often viewed as supernatural experiences that presaged future events. Building on the intellectual traditions of Aristotle and Descartes, eighteenth- and nineteenth-century philosophers started to examine dreams in an increasingly rational, secularized fashion. Soon enough, the idea that dreams originated not from otherworldly or supernatural forces, but from the dreamer's own mind, gained traction.

By the mid-1850s, medical and scientific approaches to sleep and dreams had started to flourish. One example of this growing movement appeared in 1855, when the philosophy section of the French Académie des Sciences Morales et Politiques (Academy of Moral and Political Sciences)—a learned society founded in 1795 and existing to this day—proposed a competition around the theme of sleep and dreams. The researchers posed two central questions: "What mental faculties continue, stop, or change during sleep?" and "What is the fundamental difference between dreaming and thinking?" These were great, challenging questions at the time, and they remain central questions for today's modern dream research community.

The decades that followed became increasingly rich in scientific ideas about the origins, content, and structure of dreams. Researchers were particularly interested in understanding how the mind went about constructing our nightly dream worlds. Of course, in the days before sleep laboratories, they worked largely in people's homes with participants sleeping in their own beds. Some early researchers studied their own dreams, sometimes with the help of an assistant who would monitor their sleep. Others elected to study the dreams of other people.

These researchers were not interested in studying how dreams could be interpreted, or if dreams could predict future events. Instead, they were primarily interested in determining the sources of our dreams and how to explain them. Some of these pioneers even went as far as drawing parallels between the content of people's dreams and the latest theories and discoveries about human physiology, including how the brain was believed to work. These were exciting times indeed.

Here then is a tour (in roughly chronological order) of the main ideas and observations put forth by five pre-Freudian explorers of the dream world in the second half of the nineteenth century.

❖   ❖   ❖   ❖

WERE YOU EVER TOLD that dreams last only a second? That idea, still alive and well in some circles today, can be traced to Alfred Maury (1817–1892), a professor of history and ethics at the College of France and a participant in the French Academy's contest on dreaming. Other than proposing that dreams could occur instantaneously—an idea that came to him after a particularly strange dream—Maury made several contributions to the nascent field of dream research.

In his book *Le Sommeil et les Rêves* (Sleep and dreams), first published in 1861,[6] Maury argued that our actions in dreams were guided by a mechanistic process due in part to the absence of true free will during sleep. Maury was a staunch defender of the idea of "automatism" in dreams, a view similar to how one might describe the workings of a fancy robot that is totally unaware of what it is doing and why. But he also argued that our acquired experiences—including our thoughts and knowledge about the world and all our experiences from childhood—breathed life into our dreamed actions, like a riverbank steering a fast-moving current.

Moreover, like other researchers of his time, Maury believed that our natural tendency to form associations between various daytime experiences was at the heart of how dreams were built. He argued that memories of specific thoughts, sights, sounds, events, and emotions that arose in dreams were linked through a chain of associations. But, he insisted, this process operated differently in dreams than during wakefulness. Why? Maury hypothesized that unlike the waking brain, the dreaming brain didn't work as a synchronized, coherent whole and that faculties like perception, memory, will, and judgment could all fluctuate independently of one another. As a consequence, the dreaming mind could be pulled in different directions all at once, giving rise to bizarre and incoherent dreams. Thus, Maury believed, variations in our experiences during dreams were directly tied to how different regions of the brain functioned during sleep. As we will see in later chapters, we now have ample evidence that shortfalls in cognitive abilities during dreaming (for example, our lack of self-awareness, inability to sustain focused attention, absence of logic and critical judgment) do indeed reflect varying degrees of activation across different regions of the sleeping brain. Maury would have been delighted.

Drawing on his own experiences, Maury also argued that dreams could retrieve memories long forgotten by the conscious mind, including names, places, and events. He also kept detailed notes on his nightly dreams and searched for patterns in their content in relation to factors such as the weather and the foods he ate.

Maury, however, would become best known for a series of experiments he conducted using himself as a subject. Through them, he aimed to determine if and how different sensory experiences affected his dreams. During Maury's sleep, his assistant would apply various stimuli ranging from releasing a drop of water onto his forehead to holding a bottle of eau de cologne under his nose to tickling his lips and nostrils with a feather. In most cases, Maury reported striking effects. For example, after a pair of tweezers had been held close to his ear and struck with a pair of scissors to make a light ringing sound, Maury dreamed of alarm bells ringing out a tocsin, signaling the breakout of a revolution (as he had witnessed during the Paris revolution of 1848). When a piece of heated iron was held close to him, he dreamed that robbers had broken into his house and held his feet to the fire to force him to reveal where his money was. When a burning match was held under his nose, he dreamed that he was at sea and that the powder magazine of the ship had blown up.

From these experiments, Maury concluded that not only could our senses transmit information to the brain during sleep but also that the sleeping brain would, in turn, use that information to create a relevant dream. As anyone who has ever incorporated the sound of their alarm clock into a dream knows, he was absolutely correct.

As simple as these experiments may appear to us today, they were among the first to use the scientific method, with its principle of cause and effect, to investigate dreams. Some of the earliest laboratory investigations of dreams conducted after the discovery of REM sleep in 1953 focused on the effects of external stimuli on participants' dreams; it's truly remarkable that these studies took place some hundred years after Maury had done the same.

❖ ❖ ❖ ❖

WHILE MOST PEOPLE ASSOCIATE the idea of symbolism in dreams—and especially sexual symbols—with Freud's *The Interpretation of Dreams*, the first psychologically sophisticated discussion of the nature of dream symbols was published almost forty years earlier by Karl Scherner (1825–1889) in his 1861 book, *Das Leben des Traumes* (The life of dreams).[7] Scherner wrote that people experience a weakening of the ego (self-control) during sleep so that "the activity of the soul we call fantasy is free from all the rules of reason.... It is extremely sensitive to the most delicate emotional stimuli and it immediately changes the inner life into pictures of the outer world."[8] Scherner was careful to explain that the dream did not depict objects directly; rather, it often used a different image to represent a key attribute of the object. For instance, Scherner, who had a keen interest in bodily representations in dreams, suggested that a house could symbolize the human body and that specific parts of a house could represent specific parts of the body: he described a woman who went to sleep with a terrible headache and then dreamt of a ceiling covered in cobwebs and crawling with large, nasty spiders.

Scherner was also deeply fascinated by the presence of sexual symbols in dreams, and he dedicated a dozen pages of his book to their importance. The penis, he pointed out, could be symbolized by a tobacco pipe, a knife, or a clarinet, while the female sex organ could be represented by a narrow path surrounded by houses. Sound familiar?

It's easy to see how Scherner's emphasis on the weakening of the ego during sleep and the shifty nature of dream symbols (especially those with sexual origins) influenced Freud's theory of dreams.

Although Freud himself was critical of several of Scherner's ideas, he did acknowledge that Scherner's work was "the most original and far-reaching attempt to explain dreaming as a special activity of the mind, capable of free expansion only in the state of sleep," and later even admitted that Scherner was "the true discoverer of symbolism in dreams."[9] But in the end, the world gave Freud full credit for Scherner's ideas.

❖ ❖ ❖ ❖

HAVE YOU EVER WONDERED what exactly happens to our mind as we fall asleep? Or whether we can experience things in dreams that we never experienced in waking life? Jean Marie Léon d'Hervey de Saint-Denys (1822–1892), a professor of ethnography at the College of France, tackled these and similar questions in his remarkable 1867 book, *Les Rêves et les Moyens de les Diriger: Observations Pratiques* (Dreams and the ways to guide them: Practical observations).[10] He came up with new ways to explore the incorporation of external stimuli into dreams and developed innovative techniques for inducing lucid dreams.

Saint-Denys wasn't just a passive observer of dreams; he used his finely honed skills as a lucid dreamer to investigate dreams from within, exploring their images, memory sources, and inner logic as they unfolded before (or, rather, behind) his eyes. He did so with such passion that you can't help feeling that fervor throughout his book. All in all, Saint-Denys's book is Tony's all-time favorite book on dreams.

As an only child growing up in Paris, Saint-Denys spent much of his childhood drawing and coloring. At the age of thirteen, he began recording his dreams. By the time his book was published, he had filled twenty-two volumes with meticulously detailed dream reports, many accompanied by colored drawings. One of Saint-Denys's central claims was that dream images were visual representations of ideas going through the dreamer's mind. And just as a train of thought can swiftly advance or suddenly change course, so too can dream images quickly unfurl and shift before the dreamer's eyes. Seen from this perspective, bizarreness in dreams could be explained as a natural consequence of how dream images were elicited and combined.

Saint-Denys also proposed mechanisms by which images derived from ideas and memories could meld into one another. One of his core concepts was called *abstractions*, a reference to how the mind could transfer characteristics or qualities of one person or object to another. He gave the example of how the thought of an orange could give rise to different sensory experiences in a dream, depending on whether the thought was centered on the fruit's shape, color, or smell; this would then result in dream images of a round beach ball, orange sunset, or even a grove of lemon trees. Specific qualities or details of an object, instead of the entire object, could be incorporated into a dream. Saint-Denys wrote about other forms of abstractions, including those based on wordplay, personal beliefs, moral judgment, and social traditions, and he argued that abstractions needed to be considered in order for a dream to be properly interpreted. (We'll have much more to say later about what it means to *interpret* a dream.)

Saint-Denys went on to propose a second process—one he called the *superimposition of images*—to further explain how images came to represent various ideas within a dream. He believed that when two conflicting ideas unfolded simultaneously, or when different ideas competed for visual representation within a dream, the ideas could fuse and give rise to bizarre dream elements. He gave the example of a dream in which he picked a huge peach from a tree and noticed that it looked exactly like the daughter of a friend. Saint-Denys attributed this image to an incident earlier that day, when he had heard someone mention how the girl's cheeks resembled a velvety peach. Some thirty years after the publication of *Dreams and the Ways to Guide Them*, Freud would resurrect Saint-Denys's concepts of abstractions and the superimposition of images, renaming them *displacement* and *condensation*.

As part of his many experiments on dreams, Saint-Denys added a new twist to Maury's pioneering work on the effects of sensory stimuli on dreams. He wanted to determine if, in accordance with the principle of association of ideas, specific scents could be used to evoke specific memories in dreams. (And remember, this was fifty years before Marcel Proust would famously describe the ability of the smell of a madeleine to do the same.)

To answer the question, Saint-Denys would buy a new perfume every time he traveled. Once at his destination, he would douse a handkerchief with the particular perfume and smell it daily throughout his stay. After returning home, he would wait a few months and arrange for his servant to shake a few drops of the perfume onto his pillow while he slept, but without knowing on which night his servant would do it. It worked. Saint-Denys reported multiple examples where his exposure to a given scent caused him to dream of scenes and experiences that had been associated with the perfume. Not satisfied with these findings, Saint-Denys had his servant drip two different perfumes onto his pillow, only to find elements from both trips combined into a single dream.

Following these extraordinary findings, he extended his experiments by pairing other kinds of sensory stimuli to waking events. For example, when attending a ball, Saint-Denys requested that the orchestra play a specific waltz when he danced with one woman, and a different one when he danced with a second partner. In many instances, he would then conjure up a dream image of one or the other partner when a music box of that partner's waltz was played while he slept, although the dream scenario often wasn't of a dance at all. In another experiment, he chewed on a fragrant piece of orris root while painting the statue of an attractive woman. When subsequently exposed to the root's flowery scent during his sleep, Saint-Denys dreamed of a beautiful woman resembling the one he had been painting.

Despite his numerous contributions to the study of dreams, Saint-Denys would become best known for his extraordinary talents as a lucid dreamer, capable of being aware that he was dreaming while the dream was in progress. Saint-Denys described the methods he developed to become a proficient lucid dreamer, outlining the various ways he used his awareness in dreams to explore the formation and unfolding of dreams as well as to test his memory and reasoning abilities within the dream. In one particularly amusing example, he relates a dream in which, while perfectly aware that he was dreaming, he considered Maury's idea that the brain did not work as a whole during sleep and wondered what area of the brain Maury would think was responsible for his being so clear-headed in the dream.

Much of what Saint-Denys proposed in his book can still be found in modern theories of dreaming. He described how he cured himself of recurrent nightmares using an approach that modern-day clinicians would call behavioral desensitization. And his focus on memory processes, associations of ideas, and the transformation of thoughts into mental images in dreams can be found in our own theory of dreaming described in later chapters.

Saint-Denys's *Dreams and the Ways to Guide Them* was remarkable at the time it was written and is even more so today. A free online English translation by Carolus den Blanken and Eli Meijer became available in 2016. Whether you loved Freud's *The Interpretation of Dreams* or hated it, you're sure to enjoy this remarkable book by Saint-Denys.

✧   ✧   ✧   ✧

IN APRIL 1893, there appeared in the *American Journal of Psychology* an article bearing the intriguing title "Statistics of Dreams."[11] Its author was Mary Whiton Calkins (1863–1930), a pioneering female psychologist at Wellesley College whose postgraduate training at Harvard University had been permitted only so long as it was not taken as a precedent for co-education. Undeterred by the patriarchal structure governing academia in the late nineteenth century, Calkins pushed forward with her quest for higher education and desire to establish herself as a professor and researcher. In a remarkable career spanning almost four decades, Calkins established one of the first psychology laboratories in the United States, became the first woman president of the American Psychological Association, was later elected president of the American Philosophical Association, and published four books and over a hundred papers.

"Statistics of Dreams" was among the very first of her research papers. With a novel experimental approach to the study of dreams, Calkins used statistical principles to analyze the content of almost four hundred dream reports collected over two months using herself and a thirty-two-year-old male as participants.

Much like the process used in modern laboratory studies of dreams, Calkins started her experiment by using an alarm clock to wake herself and her male subject at different times during the night. This method not only increased her chances of getting a dream report but also let her examine whether the recall and vividness of dreams changed across the night. She kept pencil, candle, and matches close at hand, since "To delay until morning the record of a dream, so vivid that one feels sure of remembering it, is usually a fatal error."[12]

By noting the time, length, and vividness of each recalled dream and charting the results, Calkins was able to show that while most dreams, including particularly vivid dreams, occurred during morning sleep, dreaming also took place during earlier parts of the night. Some seventy years later, both of these observations would be borne out by modern laboratory studies of dreams.

In an equally impressive study, Calkins showed that in nine out of ten reports, participants were able to identify evident connections between the content of their dreams and elements from their waking life. This result led Calkins to one of the key findings of her work: There is "congruity and continuity" between our waking and dreaming life. This statement foreshadowed what later became the continuity hypothesis of dreaming, which to this day remains one of the most widely and intensely studied models of dream content.

Calkins also developed standardized questionnaires that she and others could administer to large numbers of people and used them to determine the percentage of everyday dreams that contained visual, auditory, tactile, olfactory, and taste imagery. Based on her findings, Calkins proposed a hierarchy of sensory representations in dreams that would later be confirmed by modern-day home- and laboratory-based studies.

By defining key variables of interest, designing experiments that could be reproduced by others, and emphasizing quantitative over anecdotal data, Calkins's approach to dreams embodied the essence of what would become the science of dreams.

✧   ✧   ✧   ✧

OUR LAST NINETEENTH-CENTURY dream explorer is Sante de Sanctis (1862–1935), an Italian scientist who in 1899 published *I Sogni: Studi Clinici e Psicologici di un Alienista* (Dreams: Psychological and clinical studies of an alienist). Based at the University of Rome, La Sapienza, De Sanctis played a leading role in the development of psychology as a scientific discipline in Italy. Like Freud, De Sanctis believed that dreams were psychologically important and could be interpreted. Unlike Freud, however, he insisted that a true understanding of dreams could come about only by investigating dreams with a range of complimentary methods, by considering how the brain functions during sleep, and by grounding dream theories in scientific observations.

Building on the work of others before him, including that of Maury and Calkins, De Sanctis developed a multipronged approach to the study of dreams that included not only detailed questionnaires and systematic awakenings of participants during different phases of sleep but also a reliance on repeated observations and statistical analyses instead of anecdotal accounts.

Long fascinated by how dreams could reveal psychological aspects of the dreamer, De Sanctis studied the dreams of children, the elderly, criminals, patients with epilepsy, and the mentally ill, as well as healthy middle-aged adults. By examining dream content across these groups, De Sanctis came to view waking emotions as playing a vital role in how dreams were constructed. He also identified parallels and divergences between participants' waking and dreaming consciousness, documented differences in the dreams of men and women, and proposed that dream vividness was related to the development of brain functioning or, in the case of older, senile adults, to its decline.

De Sanctis also took detailed notes on the sleep of animals (including dogs and horses), convinced that such observation would help him better understand the relation between sleep and dreaming in humans. His observations of movements and twitches in sleeping animals (including the occasional bark from sleeping dogs) led him to conclude that animals also dreamed. More important, these observations also led him to consider the nature and forms of dreaming from developmental and evolutionary perspectives—two ways of conceptualizing dreams that remain at the forefront of contemporary dream research.

In his efforts to study sleep and dreaming under controlled conditions, De Sanctis became one of the first researchers to use newly developed electrophysiological instruments in his research. These included the esthesiometer (a device designed to measure mental fatigue or, in this case, sleep "depth" through the presentation of tactile stimulations of varying intensity) and the thoracic pneumograph (a band placed around the chest to measure breathing patterns).

In a particularly clever series of experiments,[13] De Sanctis showed that dreaming was less common during the deep sleep that characterized the first half of the night than it was later in the night. He found that dreams were more vivid in the lighter sleep found at the end of the night, and that dreaming was more likely to occur during periods of irregular breathing. These studies remarkably foreshadowed the formal discovery of sleep stages, including REM sleep, which occurs mostly late in the night and is accompanied by irregular breathing.[14] De Sanctis may also have anticipated several recent neuroscientific findings when he wrote about the interplay between sleep, dreams, and memory and described an intricate model of dreaming that distinguished between brain structures responsible for starting dreams and others responsible for elaborating their contents. This distinction would reappear in the 1977 activation-synthesis model of Hobson and McCarley, described in Chapter 7.

The multifaceted, integrative approach to dreams espoused by De Sanctis was perhaps best exemplified when he wrote that to be properly understood and interpreted, a dream had to be viewed as a mathematical sum: "The fundamental state of the dreamer (past experiences, intelligence, character, old habits) + the state of the moment (aspirations, passions, state of health, conditions of the organs and devices) + immediate experiences provoked by extrinsic conditions (during sleep)."[15] Today, 120 years later, we couldn't agree more.

❖ ❖ ❖ ❖

TOGETHER, THESE FIVE dream explorers from the second half of the nineteenth century gave the world a wealth of fascinating new ideas and solid scientific studies of dreaming, all published before Freud's *The Interpretation of Dreams*. Using innovative experimental methods, they tackled questions that had challenged and fascinated people for millennia, including the memory sources of dreams, the nature of dream symbols, and the role of emotional, cognitive, and physiological mechanisms in the production of dreams. But more generally, they demonstrated that it was possible to address fundamental questions about dreams empirically and scientifically, bolstering the nascent science of dreams.

But these early explorers were by no means alone. There are a dozen other authors whose writings we could just as easily have included in this section. We want to especially call out four of these works: Frank Seafield's book *The Literature and Curiosities of Dreams* (1865), F. W. Hildebrandt's *Dreams and Their Interpretation* (1875), Joseph Delboeuf's *Sleep and Dreams* (1885), and Julius Nelson's *A*

Study of Dreams (1888). These books are freely available and well worth reading.

Taken as a whole, these pioneering dream researchers helped clarify our understanding of dreams and laid the groundwork for all the scientific studies of dreams that followed. Much of what you'll find in this book is rooted in these pre-Freudian ideas and approaches to dreams and dreaming.

Chapter 3

FREUD DISCOVERED THE SECRET OF DREAMS

OR SO HE THOUGHT

LIKE MANY PEOPLE, BOB HAD ALWAYS THOUGHT THAT Freud's *The Interpretation of Dreams*, published in 1899 (the publication date was given as 1900 so it would be associated with the new century), had been an instant success, transforming Western views on dreaming almost overnight. One afternoon, while thinking about some ideas for this book, Bob turned to his father's old 11th edition of the *Encyclopedia Britannica*, published in 1910—more than a decade after Freud's *Interpretation of Dreams*—and looked up *dream*. The entry ran close to 6,000 words, not one of them about Freud's theory, although his *Interpretation of Dreams* is noted among the twenty citations of books and articles at the end of the entry. The *Encyclopedia Britannica* wasn't alone in giving Freud's work short shrift.

A decade after its initial publication, most of the major medical and psychiatric texts of the time contained little to no mention of *The Interpretation of Dreams*. Indeed, it took eight years for the initial printing of 600 copies to sell out. Worse still, as noted by Freud and subsequently detailed by others,[1] his book was poorly received by the medical, scientific, and psychiatric communities. Nine years after the publication of his book, Freud wrote that the reviews of his ideas published in scientific journals "could only lead one to suppose that my work was doomed to be sunk into complete silence."[2] And yet, *The Interpretation of Dreams* eventually became Freud's most famous work, laying the foundation for his model of psychoanalysis and shaping people's views of dreams and their relation to the unconscious for the better part of a century. What a turn of events!

When students, journalists, or strangers ("Hey, there's a dream researcher at the party!") ask Tony about his take on Freudian dream theory, he often begins by asking them what they thought Freud's new ideas about dreams were. The replies, often extraordinarily short, invariably include at least one of the following ideas: dreams arise from the unconscious; dreams are really about sex; dreams are related to repressed wishes; dreams are symbolic and, to be properly understood, need to be interpreted. But as you saw in Chapter 2, most of these ideas had been articulated well before *The Interpretation of Dreams*. Clearly, Freud's ideas were not all as novel as most people now believe.

THE DREAM AS GUARDIAN OF SLEEP

Freud's theory was the first to propose that dreams had two interrelated functions. One was to give expression to repressed wishes of a sexual (or at times aggressive) nature, often dating from early childhood. The other function, not as well known, was to protect sleep from being disturbed. "Dreams," Freud explained, "are the guardians of sleep."[3] Here's how it worked.

Freud postulated the existence of a *censor*, a sentinel-like mechanism within the mind that kept unacceptable unconscious material from ever reaching conscious awareness during the day. During sleep, however, the sentinel became ineffective. It let down its guard, so to speak, allowing this unacceptable material to rise toward consciousness. Because repressed wishes were by their very nature immoral and antisocial (according to Freud), it was critical to keep them from being expressed directly even during sleep, as this would shock the dreamer into wakefulness. A residual dream censor (known as the dreamwork) was therefore tasked with distorting the normally repressed unconscious material into unrecognizable form. Freud proposed four disguise mechanisms—condensation, displacement, considerations of representation, and secondary revision—which together carried out the dreamwork. Dreams thus allowed the partial expression of repressed, often prurient, wishes (dreams as "wish fulfillment") while ensuring the preservation of restful sleep (dreams as the "guardians of sleep").

It bears noting that Freud was unyielding in his belief that *every* dream was an attempt at wish fulfillment. In other words, no dream could materialize unless first infused with the psychic energy of a repressed wish.

This conceptualization of dreams gave rise to a key distinction between the dream's manifest content—the actual dream as experienced and reported by the dreamer—and its latent content—the "true" meaning of the dream, which was the repressed wish that "energized" the dream. A dream's hidden meaning could be revealed through the technique of free association, which consists of having the dreamer provide an uncensored description of the feelings and thoughts evoked by various elements in the dream. According to Freud, these free associations, in the hands of an experienced psychoanalyst, could be used to "undo" the distortions created by the dream censor, thereby tracing the dream back to the unconscious conflicts and desires that gave rise to its manifest content.

Although the lay public of his time was generally excited by Freud's description of the origins and meaning of dreams, the same could not be said for clinicians, philosophers, and scientists, who expressed numerous criticisms. Here are just a few of the objections raised in the years following the publication of *The Interpretation of Dreams*:

+ Freud's focus on repressed infantile wishes as the primary source of dreaming was excessively restrictive; dreams could emanate from a range of innate reflexes, drives, and emotions and, in many cases, could be shown to result from the dreamer's preoccupations with daily life with no need to invoke stealthily concealed impulses.
+ His reliance on anecdotal evidence, selected case reports, and speculative assumptions rendered his dream theory anything but scientific.
+ He unduly minimized the importance of the manifest content of dreams.
+ Contrary to his theory, it was well established that many dreams—nightmares in particular—did not protect sleep, but instead woke up the dreamer.
+ The results derived with his method of dream interpretation were often biased and forced, and his conclusions arbitrary.
+ Many types of dreams, most notably nightmares, showed no clear evidence of censorship or distortion.
+ Perhaps most critically, his theory failed to meet the most fundamental requirements for any scientific theory—it was neither testable nor falsifiable.

Doubts about the rigor and legitimacy of Freud's dream theory weren't limited to those outside of Freud's inner circle. Some of the most animated disagreements came from his own followers, including Alfred Adler (one of Freud's first disciples, who would go on to establish the field of personal psychology) and Freud's heir apparent, Carl Jung.

Jung and the Alternative Clinical Conceptualizations of Dreams

Freud saw dreaming as a relief valve for pathological desires. Jung, on the other hand, believed that dreams played a vital compensatory role in the development of a person's personality by presenting the dreamer with unconscious material that needed to be recognized (and integrated) for the dreamer to achieve a more balanced sense of self. This material, according to Jung, often arose from the dreamer's personal unconscious and could include "the meanings of daily situations which we have overlooked, or conclusions we have failed to draw, or affects we have not permitted, or criticisms we have spared ourselves."[4]

Jung also believed that dreams could emanate from what he called the collective unconscious, a deep stratum of the unconscious that was common to all humankind and encompassed the accumulated experience of the human species. He postulated that this ancestral aspect of personality was genetically inherited and that it expressed itself in the form of archetypes (universal patterns and images) that could be observed cross-culturally in fairy tales, myths, sacred rituals, mystical experiences, many works of art, and of course in people's dreams. Jung described a wide range of archetypal motifs and symbols, including such archetypal figures as the shadow, the trickster, the wise old man, the great mother, and the hero.

Jung also believed that dreams could have an anticipatory or "prospective" function. By tracking the dreamer's past, the unconscious processes underlying dream formation could present the individual with visions of probable situations and challenges, unmapped potentials, or imaginable results that lay in the dreamer's future. As we will see in Chapter 8, these notions are not as far-fetched as they may seem.

Thus, while Freud wrote that dreams were "abnormal psychical phenomena" akin to neurotic symptoms, and while he emphasized their deceitful nature, Jung focused on dreaming as a wholesome, natural process and highlighted its creative, transcendent, and at times problem-solving nature. Like Freud, Jung proposed techniques for working with dreams and was convinced that dream interpretation could lead to important personal insights. But he also recognized the sometimes arbitrary nature of dream interpretation and, in contrast to Freud's dogmatic assertion of the importance of his theory, Jung doubted that his own techniques even deserved the name "method"; he was anything but dogmatic about their use.

The dream theories of Freud and Jung paved the way during the 1900s for a number of other, lesser-known clinical conceptualizations of dreams. These include the view of Alfred Adler, mentioned earlier, that the manifest content of dreams was intimately tied to the dreamer's waking concerns and lifestyle, not to the unconscious as proposed by his mentor; the existential-phenomenological approach to dreams developed by Swiss psychoanalyst Medard Boss that considers the dream as an authentic experience of "being-in-the world" as real as any waking experience; the focal conflict theory of Thomas French and Erika Fromm, two classically trained psychoanalysts who proposed that dreaming reflects the ego's attempt to solve vital concerns in the dreamer's waking life; and Frederick Perls's Gestalt-based approach in which different dream elements are understood as projections of both accepted and disowned aspects of the dreamer's personality. Beyond these, the wave of experimental dream research that followed the discovery of REM sleep in 1953 gave rise to dozens, if not hundreds, of other theories about the nature and function of dreams in the second half of the twentieth century.

Where does all this leave Freud's claims that "dreams are the fulfillments of wishes" and that "the dream is the guardian of sleep?" Innumerable dream studies have led to the simple conclusion that there is little or no empirical support for either of these proposed dream functions. Moreover, the overwhelming majority of sleep and dream scientists have long abandoned Freudian conceptualizations of dreams in favor of more parsimonious and testable models rooted in modern clinical and neuroscientific research. This is not to say that modern-day researchers have abandoned the notions that dreams can carry personal meanings, reflect the dreamer's current waking concerns, and reference long-past memories, or that working with dreams can be clinically useful. All of these concepts have been, and continue to be, the subject of innovative studies. They just have very little to do with actual Freudian dream theory.

By now some of you may be wondering—given that so many of Freud's contemporaries thought his ideas about dreams were wrong, that a wealth of alternative dream theories arose over the next hundred years, and that there was so little empirical evidence to support the Freudian dream model—how did his theory become so ingrained in Western culture? Therein lies something of a saga.

Because Freud had made *The Interpretation of Dreams* the cornerstone of his psychoanalytic theory, it became virtually impossible to criticize his dream theory without calling into question the whole psychoanalytic enterprise. As a result, disagreements about the presumed functions of dreams were invariably diverted into quarrels about a host of other matters, such as the concept of repression, the nature of human memory, the origin of neurotic symptoms, models of childhood development, the clinical merits of free association, and the makeup of the unconscious as well as its presumed influence on everyday life. To some extent, this state of affairs still holds true today, and readers unfamiliar with the extreme virulence (some would say religious fanaticism) of these fights are invited to read up on the "Freud wars."[5] Even a cursory search will lead interested readers into some fascinating but decidedly poisonous literature.

Despite these battles, the first seventy-five years of the twentieth century were good to Freud's theory. Backed by a multitude of highly motivated proponents and a wealth of establishments dedicated to the training of mental health professionals in the methods of psychoanalysis, the psychoanalytic school of thought slowly morphed into a movement so commanding and widespread that many of its tenets became omnipresent in the fields of medicine, psychiatry, and clinical psychology. The movement also came to permeate the social sciences and, perhaps more important, the arts. From departments of history and literature to the paintings of Salvador Dalí to Alfred Hitchcock's *Spellbound* and the works of countless writers, Freudian-laced ideas about dreaming and the mind filled the media and the arts, capturing the imagination of young and old alike. As aptly put by University of California, Berkeley, psychologist John Kihlstrom, "More than Einstein or Watson and Crick, more than Hitler or Lenin, Roosevelt, or Kennedy, more than Picasso, Eliot, or Stravinsky, more than the Beatles or Bob Dylan, Freud's influence on modern culture has been profound and long-lasting."[6] And when it comes to dreams, Freud's legacy is without equal.

But as we've seen, people confuse the myths surrounding Freud and his ideas with what he actually wrote. Many of us mistakenly believe that Freud was the first to suggest that dreams contained repressed wishes or desires, or that dreams were symbolic, or that they came from the unconscious. The concept of the unconscious, which is typically ascribed to Freud, can be traced back through millennia; the word itself was coined some hundred years before Freud was born.[7] And even the distinction of putting forth the first clinically informed theory of the unconscious mind belongs not to Freud, but to the French psychiatrist Pierre Janet, whose writings greatly contributed to Freud's own psychoanalytic thinking. Still, it's now 120 years after the publication of *The Interpretation of Dreams*, and most of these ideas remain strongly—if not unquestioningly—associated with Freud.

Finally, as with all good marketing efforts, timing was important for Freud. His psychoanalytic dream theory arrived at a time when popular views of dreaming had become more rational and secularized, describing dreams as meaningless nocturnal events that could be explained by natural bodily processes. But like people today, large segments of the population rejected this notion and remained convinced that dreams, however bizarre, carried important messages, and ones that called for interpretation. This basic idea, old as humankind itself, is exactly what Freud brought back. With a masterful and compelling writing style, Freud managed to weave ideas and findings from a wide range of sources and disciplines into a rich, absorbing narrative that not only told people they were right to believe that their dreams were important but also showed them how and why. And, truth be told, there's something narcissistically comforting for all of us in the idea that deep down, we're really unknowable beings—that our behaviors are guided by motives and desires we remain largely unaware of, and that our dreams can reveal the very essence of who we are.

MINDS, NEURONS, AND EEL TESTICLES

Freud's path to his theories of dream interpretation and psychoanalysis was not a straight one. Today, few people know that he originally trained as a neurologist and not, as many naturally assume, a psychiatrist. Even fewer know that his first research project focused on the sexual organs of the eel. Debates about the eel's mysterious reproductive habits date back to Aristotle's time. In an attempt to answer this age-old question, the young Freud spent weeks dissecting hundreds of eels in a painstaking but ultimately fruitless search for their elusive male gonads. "All the eels I have cut open," he wrote to a boyhood friend, "are of the tenderer sex."[8] We can only imagine how this effort may have affected his dreams, not to mention his dream theory. He was only nineteen.

Freud then spent six years working at the Institute of Physiology under the supervision of Ernst Wilhelm von Brücke, a famous German physiologist whom Freud would later describe as "the greatest authority I have ever met." During that time, Freud published the first report detailing the structure and function of a part of the brainstem known as the medulla oblongata and developed a new staining procedure to highlight nerve cells in dissected tissues.

A dozen years later, Freud had begun to conceptualize his psychoanalytic theory, and as he wrote to his close friend and confidant Wilhelm Fliess, his desire to present a neurologically accurate model of the psyche became an all-consuming affair. Freud started working on a project that would use ideas and breakthroughs in neurology to explain normal and abnormal mental processes. An important portion of the resulting manuscript was dedicated to sleep and dreams. It was during this period that Freud experienced his now-famous dream of "Irma's injection," the dream that would singlehandedly lead him to the conclusion that dreams were the fulfillment of wishes. Later, in a letter to Fliess, Freud wondered, "Do you suppose that someday one will read on a marble tablet on this house: 'Here, on July 24, 1895, the secret of the dream revealed itself to Dr. Sigm. Freud?'"[9]

In this little-known manuscript, drafted four years before publication of *The Interpretation of Dreams*, Freud first mentioned his wish-fulfillment hypothesis of dreams. But he eventually abandoned the project and never referred to it in any of his subsequent published works. It would be another fifty-five years, long after Freud's death, before his unfinished manuscript was discovered and published in 1950 under the title *Project for a Scientific Psychology*.[10]

Ironically, the story doesn't end here. In a remarkable twist of fate, another long-lost manuscript, this one from one of Freud's contemporaries, would also resurface decades after it had been written. This manuscript, recovered in 2014, belonged to Santiago Ramón y Cajal, a Spanish histologist and anatomist who won the 1906 Nobel Prize in Physiology or Medicine for his discovery of nerve cells. Thanks to his revolutionary work and ideas, Ramón y Cajal is widely considered to be the father of modern neuroscience. His lost manuscript, however, was not another essay on anatomy or histologic techniques. It was about dreams—*his* dreams. From 1918 until his death in 1934, Cajal had kept a dream journal with the avowed purpose of proving Freud wrong.

Cajal had always been open about his disagreements with Freud's view of the mind. But his disdain for Freud's theory of dreams was never made as explicit as when he wrote to a friend that "except in extremely rare cases, it is impossible to verify the doctrine of the surly and somewhat egotistical Viennese author, who has always seemed more preoccupied with founding a sensational theory than with the desire to austerely serve the cause of scientific theory."[11]

Just as Freud's training in neurophysiology will come as a surprise to most, few are aware that Cajal's earliest scientific interests lay in experimental psychology, including the study of suggestion, hypnosis, and the mechanisms underlying sleep. And of Cajal's 350 publications, only one deals with dreams. Published in 1908, it opens with the following sentence: "Dreaming is one of the most interesting and most wondrous phenomena of brain physiology." How can we disagree? Cajal intimated that he would publish a detailed work on "sleep and the phenomena of dreaming," in which he would summarize "thousands of self-observations contradicting Freud's theory." Presaging a neuroscientific model of dreams that would be rendered famous almost seventy years later by Harvard psychiatrists Allan Hobson and Robert McCarley, Cajal held that dreams resulted from wild neural firings in various regions of the brain. Though Cajal never completed his promised work, his dream diary and accompanying notes have recently been published in a wonderful book[12] that not only examines Cajal's life work but also reveals what these dreams tell us about one of the greatest scientists of the nineteenth century.

It is mind-boggling to realize that the man known as the father of neuroscience, a neurophysiologist who also worked as an experimental psychologist, spent sixteen years recording his dreams in order to debunk the theories of the man known as the father of psychoanalysis, who first worked as a neurophysiologist before shifting his focus to psychology. But then again, such is the spell-binding nature of this "most wondrous" phenomenon we call dreams.

Chapter 4

THE BIRTH OF A NEW SCIENCE OF DREAMING

OPENING WINDOWS UNTO THE SLEEPING MIND

EVERYTHING WE THOUGHT WE KNEW ABOUT DREAM-ing began to change one cold December night in 1951, when Eugene Aserinsky thought his son Armond had woken up. Aserinsky, your classic starving graduate student, was enrolled at the University of Chicago. With a pregnant wife and young son at home, life was hard. His daughter later recalled, "We were so poor my father once stole some potatoes, so we would have something to eat." But the thirty-year-old Aserinsky, who had talked his way into graduate school even though he didn't have a bachelor's degree, was driven to get his PhD. And on that December night, he was once again trying to record the tiny electrical signals produced by the eye movements of his sleeping eight-year-old son.

Aserinsky was lucky because Frank Offner had preceded him at the University of Chicago, where Offner had invented the Offner Dyno-graph. This device recorded electrical signals, like those generated by eye movement, on a continuous sheet of folded paper. When it was working, the Dynograph would let Aserinsky detect when Armond was blinking and obviously awake. It was just such bursts of electrical signals, drawn on the Dynogram paper sometime after Armond had fallen asleep, that led Aserinsky to assume his son had awoken. He went into the bedroom to see how he was doing.

But Armond wasn't awake. He was sound asleep! Something was obviously wrong with the Dynograph. Except there wasn't; instead, something was wrong with Aserinsky's understanding of sleep.

Less than two years later, Aserinsky and his graduate school advisor, Nathaniel Kleitman, published a two-page article in the prestigious journal *Science*. The paper, "Regularly Occurring Periods of Eye Motility, and Concomitant Phenomena, during Sleep,"[1] reported the presence of periods of rapid, jerky eye movements occurring periodically throughout the night. In one fell swoop, Aserinsky and Kleitman had discovered rapid eye movement (REM) sleep and its recurring appearance every 90 minutes across the night.

Four years later, in 1957, Kleitman published another paper, this one with another student, William Dement, whom Aserinsky had told of an apparent connection between REM sleep and dreaming. Curious about this possibility, Dement recorded the sleep of nine adults for a total of 61 nights, awakening them on average six times a night and collecting a total of 351 dream reports. When he compared the reports from REM awakenings to those from nonREM, the results were stunning.[2] When participants were awakened from nonREM sleep, they provided a "coherent, fairly detailed description of dream content" only 7 percent of the time. In contrast, after being awakened from REM sleep, they provided coherent reports 80 percent of the time—more than ten times as often. No longer was dreaming just a mystical mental phenomenon that seemed to come from nowhere except maybe the hidden recesses of our psyche. Suddenly, there was a *biology* of dreaming.

These newly discovered periods of "rapid, jerky eye movements," as Aserinsky and Kleitman called them, were not just related to dreaming. We now know that the brain's normally careful regulation of a host of bodily functions seems to go offline during REM sleep. Heart rate, blood pressure, and breathing all vary widely during REM sleep. Not only that, but men have prolonged erections during REM sleep, and women have a similar swelling of the clitoris. There are also pronounced changes in brain activity during REM sleep; electrical activity in the brain becomes indistinguishable from that of a person who is awake. Other changes leave us effectively paralyzed during REM sleep, with a near complete loss of muscle tone. Even the chemicals being released in the brain and regulating its activity change during REM sleep. In short, REM sleep is a unique brain and body state that is not seen at any other time of the day or night.

Aserinsky's discovery came just twenty-two years after Hans Berger, a professor of neurology and psychiatry at the University of Jena in Germany, reported the first recording of the human electroencephalogram (EEG). An electroencephalogram is a recording ("gram," as in *telegram*) of the electrical activity ("electro") in the brain ("encephala," from the Greek *en* + *kephale*, meaning "in the head"). Berger was really interested in the physiological basis of psychic energy,[3] but when those studies went nowhere, he had turned to studying the electrical activity of the brain. His timing couldn't have been better for the discovery of the unique pattern of brain activity seen in REM sleep.

But the cyclic recurrence of other characteristics of REM sleep could have been discovered at any time in human history. Every adolescent boy knows about the erections that occur during sleep, and the Roman physician Galen (who was personal physician to the emperor Marcus Aurelius) described erections during sleep way back in the first century.[4] He even noted their correlation with dreaming. If someone had bothered to chart their occurrences across a full night of sleep, the periodicity of REM sleep and its association with dreaming would have been discovered two thousand years earlier. Today, new parents often notice the rapid eye movements visible through the thin eyelids of their sleeping babies. REM sleep was there to be discovered, if anyone had thought to look.

THE REM CYCLE

Although researchers learn more about the different phases of nocturnal sleep every year, the basics are now clear. Humans cycle in and out of REM sleep about every 90 minutes, all night long. Occurring between these REM periods are three distinguishable stages of non-REM sleep, called N1, N2, and N3, each representing deeper sleep than the one before. A good night of sleep looks something like the hypnogram in Figure 4.1.

The thicker lines representing REM sleep appear regularly every hour and a half, and the first REM period occurs just after midnight. This 90-minute period remains relatively constant across the night, but the length of REM periods increases with each cycle. The amount of time spent in deep N3 sleep decreases, and it can disappear altogether by the middle of the night. These sleep stages were originally characterized in a scoring manual written by Allan Rechtschaffen and Anthony Kales in 1968, who brought together a group of experts and allegedly wouldn't let them leave the final day's meeting until agreement had been reached. In the end, the group defined five stages—

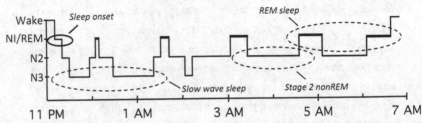

Figure 4.1. A good night's sleep, showing N1, N2, and N3 nonREM sleep and REM sleep.

REM and stages 1–4 of nonREM.[5] Some forty years later, nonREM stages 3 and 4 (together known as slow-wave sleep) were combined, giving us the current nomenclature of REM and N1–N3.

When sleep researchers score a sleep record, they don't look just at the EEG; they also look at recordings of eye movements and muscle tone, which are called the electrooculogram (EOG) and electromyogram (EMG). When you take these names apart, you can see their similarity to the name of the electroencephalogram (EEG). The electro-*encephalo*-gram, electro-*oculo*-gram, and electro-*myo*-gram are simply recordings of the electrical activity of the brain, the eyes, and the muscles, respectively. As mentioned earlier, all of these activities are altered in REM sleep. But they also differ between the three stages of nonREM sleep. In the top panel of Figure 4.2, you can see the EEG recordings of brain activity in wake, N2, N3, and REM. The differences are dramatic.

When you're awake, there's not much to see in the EEG. Of course, this obviously doesn't mean that nothing is happening. It's just that there is little consistency in how different nerve cells, or neurons, in the brain are behaving. When we record the electrical activity of your brain with an EEG, we stick two electrodes onto your scalp and merely record how the voltage between them changes over time. You could do this by taking a simple voltmeter and holding its two probes against your scalp, just like you'd hold them against the positive and negative poles of a battery to check its strength. Holding the probes against your head, you could watch the needle rushing back and forth. What you would see is your EEG.

Well, this isn't exactly the case, since that voltmeter you found in the basement is measuring from 1 to 6 volts, while the biggest signal in your EEG is about 100 millionths of a volt, or 100 microvolts. Still, the principle is exactly the same.

What is actually getting measured? Imagine standing outside a stadium before the start of a football game and holding a stethoscope against the stadium's concrete wall. What you'd hear is a con-

Sleep Physiology

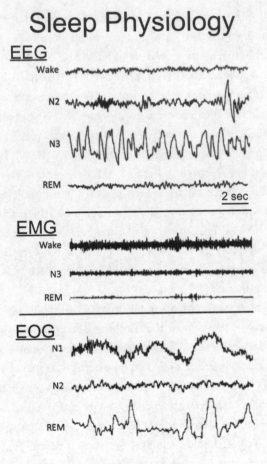

Figure 4.2. EEG, EMG, and EOG recordings
from different sleep stages.

stant low roar, the sound of thousands or tens of thousands of people each engaging in their own personal conversations. If you recorded the intensity of the sound, it would look like the Wake EEG in our figure—just a lot of little fluctuations as the number of people talking happened to go up and down. Now imagine that the game has started, and you're between plays. Players are shifting around, and the sound intensity might look like the recording from N2 sleep. There are many larger perturbations in the sound intensity as the fans respond to players coming out of the huddle and settling into position. You can see the snap of the ball near the end of the recording, when everyone shouts at once. The N3 recording is exactly what you'd see if the fans all started clapping in unison. While you could never hear one person clapping from outside the stadium, when thousands of them all clap in unison, you could probably hear it even without your stethoscope.

Bob lives a little over a mile from Harvard's football stadium, and when the wind is out of the west, he likes to sit in his backyard and listen to the roar of the fans when Harvard scores. Similarly, when we see an EEG pattern like that shown here for N3 sleep, we know that large numbers of neurons—not thousands or tens of thousands, but millions or tens of millions—are all firing rhythmically together, once or twice every second. As we go from being awake into deeper and deeper stages of sleep, the EEG signal gets stronger, with more and more cells alternating between firing synchronously and resting for a few tenths of a second, as the entire brain focuses on reviewing and replaying our memories from the day. The large, slow waves of N3 sleep have led to its becoming known more colloquially as slow-wave sleep.

But what's going on in REM sleep? The EEG looks just like wakefulness. That's partly why Aserinsky thought his son Armond was awake. In fact, for many years REM sleep was also known as paradoxical sleep, because the EEG seemed to be saying you were wide awake when, in fact, you were sound asleep. It's why sleep researchers need the EMG and EOG as well. You can easily distinguish REM sleep from wake if you look at the muscle tone, shown in the EMG (see Figure 4.2). As you might expect, your body relaxes when you first fall asleep and then relaxes more and more as you sink into deeper and deeper stages of nonREM sleep and your brain waves get larger and slower. But a strange thing happens when you go into REM sleep. Your brain waves speed up again as if you were awake, but your muscle tone drops to near zero. In fact, if you were sitting up in a chair, you'd likely fall off it. You are in a state of atonia, meaning you have no muscle tone and no ability to control your muscles. The combination of a wake-like EEG and a flat EMG is a dead giveaway that you're in REM sleep.

Sometimes REM atonia appears in people while they're awake, and the results are quite striking. We mentioned narcolepsy back in Chapter 1, when we were talking about people who couldn't distinguish their memories of waking experiences from those of their dreams. Narcolepsy is a sleep disorder in which the brain circuits that control the wake-sleep cycle are disrupted, causing sufferers to go into REM sleep as soon as they fall asleep, instead of only after an hour or two of sleep. In addition, the paralysis, normally only seen during REM sleep, can creep into wakefulness, resulting in attacks of cataplexy (from the Greek—*kata* meaning "down" + *plessien* meaning "strike") during which the person collapses to the ground as if "struck down" by an invisible hand. Curiously, these attacks are most frequently triggered by strong emotions and most commonly by laughter. Although awake, the person has lost all muscle control and might remain like this for up to a minute before muscle tone returns. You can find videos online of people with narcolepsy having cataplectic attacks. If you didn't know they had narcolepsy, it would be difficult to imagine what was happening to them.

A more common example of REM sleep atonia creeping into wakefulness is sleep paralysis. Sleep paralysis can occur when you wake up out of REM sleep, usually in the morning. As your brain shifts you from REM sleep to wake, the system producing REM sleep atonia can be slow in switching off, causing you to wake up still paralyzed. To make matters worse, your brain also seems to want to continue dreaming, even though you're awake and your eyes are open. The result is a visual hallucination integrated with the image of your bedroom provided by your eyes. People see strangers or even monsters coming into their room. A friend saw a giant spider, at least two feet tall, hanging in the corner of her bedroom. About a quarter of adults have had such an experience at some point in their life, most commonly after not getting much sleep for several nights in a row.

The hallmark of REM sleep, from which it gets its name, is its rapid eye movements. (Also known by its initials as R.E.M. sleep, it is the only sleep stage with a rock band named after it.) These bursts of rapid eye movements (or REMs) are evident in the EOG shown in Figure 4.2. During these bursts, your eyes jerk quickly from left to right, moving as quickly as when you intentionally shift your gaze to catch some sudden movement on one side or the other. They'll continue leaping back and forth for a couple of seconds and then stop, only to start again anywhere from several seconds later, as in this example, to a minute or two later, until the REM period finally ends.

REMs are triggered deep in the brainstem, in one of the brain regions critical for the control of REM sleep. Although it remains controversial, the dominant theory as to why we have REMs is the "scanning hypothesis." First proposed by Howard Roffwarg in 1962,[6] shortly after the discovery of REM sleep, the scanning hypothesis posits that the rapid eye movements of REM sleep are caused by the brain tracking the action in our dreams with our gaze direction. Another possibility is the opposite of the scanning hypothesis; namely, that movement in the dream is guided by the brain's attempt to produce a narrative that matches the movement of the dreaming person's eyes.

What makes REM sleep such a unique brain state is that these three distinctive features—rapid eye movements, atonia, and a wake-like EEG—all appear within about 30 seconds of one another and then remain together for up to half an hour before quietly disappearing as quickly as they had come.

Although REM sleep is dramatically different from nonREM, the differences between the three phases of nonREM are more a matter of degree. The N3 stage shows many more slow waves (markers of "deeper" sleep) like those seen in Figure 4.2, but otherwise is similar to N2. N1 sleep usually lasts for only a minute or two right after you

fall asleep, but it has one unique feature—namely, the slow eye movement (SEM), seen in this figure. During these slow, rolling eye movements, the eyes roll back and forth at less than a tenth the speed of
rapid eye movements,[7] taking 2 to 4 seconds to move back and forth.
The brain mechanisms producing SEMs and the possible function
of SEMs remain unknown, but their appearance is tightly linked to
the time when your awareness of the real world fades away. Indeed,
the hallucinatory imagery that often accompanies sleep onset usually
begins within seconds of the appearance of SEMs.[8] Sleep researchers
don't yet know whether these two phenomena are meaningfully connected to each other, but we are certain that both N1 and REM sleep
are characterized by this combination of stereotypical eye movements
and hallucinatory imagery.

DREAMING ACROSS THE NIGHT

When exactly do people dream? We dream at sleep onset and in
REM, but do we dream during N2 and N3 as well? Knowing when
we dream helps us understand which brain mechanisms are involved.
Moreover, how much of the night are we actually dreaming? Freud
had thought we dream very little, and maybe only if we have neuro-
ses. If we dream all night long, his theory is hard to support. And are
our dreams different, depending on what sleep stage we're in, or what
time it is? The quick and dirty answers are that we dream in all stages
of sleep; we probably are dreaming most of the night, but more consistently in some stages and at some times than others; and, on average,
our dreams are different from one sleep stage to another as well as
from early to late in the night.

More complete answers are, predictably, more complicated. For
a start, we have to go back to Chapter 1, to our discussion of what
counts as a dream. In that chapter we saw that there is no agreed-upon
definition of what a dream is, and how much and when we dream will

depend on our definition. By one definition, we have to include day-dreams, and by another we might only count more complex dreams from REM sleep. To avoid all of these arguments, we'll use a simple definition. We will count as a dream any mental experience that occurs during sleep—any thoughts, feelings, or images that come into our awareness while we sleep. By images, we don't just mean visual images. We mean any sensations: those that arise internally, like the sensation of a sore muscle or a stomachache, and those that occur externally, including sights, sounds, and smells, or tastes, touch, temperature, body position, and balance. The sensations can be true sensations, as of a full bladder, or hallucinatory, such as an image of a face or the sound of a trumpet. A dream can be a complex, bizarre story with sights and sounds, people, animals, traffic, arguments, confusions, joys, and fears. Seen this way, human dream experiences lie on a continuum ranging from isolated sensations or thoughts to epic otherworldly journeys. All of these experiences are examples of sleep mentation and fall into our definition of *dream* from Chapter 1.

Given this working definition, do we dream in nonREM sleep? Absolutely. As our definition of dreaming has evolved over the years, the number of REM awakenings that produce reports of dreaming has consistently hovered around 80 percent. But for nonREM N2 awakenings, the number has slowly risen from 7 percent in the original paper of Dement and Kleitman to a best guess of 50 to 60 percent,[5] and some studies have reported rates of over 70 percent. Even higher rates are seen for "hypnagogic" dreams, collected from nonREM N1, in the first minutes of sleep. The word *hypnagogic* comes from the Greek words for sleep (*hypnos*) and leading (*agōgos*) and refers to the sleep-onset period that leads us into sleep. People awakened from this hypnagogic period report dreams about 75 percent of the time, a rate indistinguishable from the 80 percent of REM sleep. Even studies of deep N3 sleep obtain dream reports close to 50 percent of the time. In the end, all stages of nonREM sleep yield dream reports the majority of the time.

Some researchers have even suggested that dreaming may occur all night long, whether or not we can remember anything upon awakening. This possibility is supported by numerous anecdotal reports of delayed recall. For example, you might wake up not remembering having dreamt, but then some subsequent event—getting into the shower or seeing a cat run into the street—brings up a sudden vivid recollection of the details of a dream involving the shower or a cat. The absence of dream recall clearly is not proof of the absence of dreaming.

Of course, this doesn't mean that our dreams look the same in all stages of sleep. For example, if you count how many words are in a dream report, N1 reports are shorter than N2 reports, which in turn are shorter than REM reports. What this means isn't as obvious as you might think. Longer dream reports from REM sleep might indicate that dreams actually last longer in REM, or perhaps they just take more words to describe—maybe because they're more vivid or more bizarre. But it could also mean that we just remember more of our dream when we awaken from REM sleep. Most likely, all three of these explanations apply.

TALKING ABOUT DREAM CONTENT

Debbie came into the sleep lab for her first of three nights in our dream study. She brought her pajamas and toothbrush. After changing, she sat patiently while we spent a half hour attaching electrodes to her head to record her EEG, next to her eyes for the EOG, and on her chin for her EMG. The electrodes would constantly monitor her brain waves, eye movements, and muscle tone across the night, and we would record them all compactly on our computer hard drive—as opposed to the thousand pages of 8" x 12" fan-folded paper used as recently as the 1990s. While wiring her up, we explained what we were looking for. "In this study we're interested in the thoughts you have within your dreams. Normally, people don't report that they were thinking anything. It seems that the visual images, actions, and emotions in a dream are so intense that they dominate the report. So, when we wake you later tonight for a dream report, we want you to pause for a few seconds and try to remember as much detail as possible from any dream you might have been having right before we woke you up. Then, when you start to report it, start with any thoughts you remember. If you don't remember any thoughts, that's perfectly fine. And, of course, if you don't remember a dream, or if you feel uncomfortable reporting its content, just say so. We'll be recording what you say so we can review it later." Once wired up, Debbie got into bed, and within 10 minutes was sound asleep. Ninety minutes later, she was awakened by a recording asking her to report any dream that she could recall "from the period preceding your awakening."

The method we just described is one of many ways modern scientists collect dream reports. Sometimes we have people write down their dream reports; at other times we record them. Sometimes we have people use this method at home; and sometimes, as in Debbie's case, we bring them into the sleep laboratory. When study participants are at home, we might ask them to report their dreams every time they wake up, or just in the morning, or just when they feel like it. In the lab, we have them report every time we wake them up—in some studies as many as a dozen times in a night—and then again in the morning, when they wake up on their own. Sometimes we even have participants report any dreams they can remember from their past—in the last week, the last month, the last year, or in their entire life. Sometimes we want only their most memorable dreams; at other times we want the most recent. It all depends on the question we're trying to answer.

And then there's what we want our participants to report. This can be as little as choosing between three alternatives: not remembering they were dreaming; remembering that they were dreaming, but not recalling any content (known as a "white dream"); or remembering both that they were dreaming and what they were dreaming. More commonly, we ask participants to report "everything that was going on in your mind before you woke up—everything you saw, heard, smelled, and did, everything you thought and everything you felt. Don't include what you think it meant or where you think it came from. Just report the dream." Or to paraphrase the TV detective Joe Friday, "Just the dream, ma'am, just the dream."

Sometimes we use "affirmative probes" to get information about a particular question we're interested in. For example, there are very few reports of smells or tastes in dreams, so maybe we don't dream about these much. But Bob did a study in which he used a pager (this was before the time of cell phones) to beep participants during the day and asked them to report everything they had been experiencing just before they were beeped. Altogether, he managed to collect a couple hundred reports while participants were awake and eating. But although many people described what they were doing, seeing, and hearing while they ate, almost none reported tasting or smelling anything. It turns out that we just don't report that kind of information. Thus, in a dream study, we might add the instruction, "Pay particular attention to reporting any smells or tastes that you experienced during the dream."

In his 1988 book *The Dreaming Brain*,[10] the then Harvard sleep and dream researcher Allan Hobson described "uncertainties" as one of the types of bizarreness found in dreams. In one report, a participant said, "I was sitting by the ocean, or maybe it was a swimming pool." What does this mean? Can't she remember which it was? Some years later, Bob did a study with Hobson using affirmative probes to answer this question. They asked participants to "underline any part of your report that you're uncertain about. Then indicate whether the uncertainty is due to your having forgotten the details of the dream, or whether the uncertainty was actually present in the dream. Finally, in the latter case, indicate whether you were aware of the uncertainty while you were dreaming, or only after you woke up." Affirmative probes have helped dream researchers to resolve numerous questions like this about the detailed structure of dreams.

We can also ask participants to fill out questionnaires after each dream (much like Mary Calkins did back in the nineteenth century). We might give them a list of emotions and ask them to check off all the ones they experienced during the dream, or ask them to list the characters in the dream and indicate whether they were famous people, people they knew personally, or people they did not know at all. We might ask participants to identify the dominant emotion of the dream, or how long it seemed to last. Again, the information we ask for depends on the question we're trying to answer.

Once we've collected a batch of dream reports, what we do with them depends on why we collected them. In science, there's always an underlying question that we're trying to answer. The examples of affirmative probes that we just described were all based on specific questions we were trying to answer.

In one study, Bob wanted to find out if memories of the day's events were actually replayed in dreams, as they would be if we recalled the event the next day. He had participants write down their dreams and then underline any part of the report for which they felt they knew the waking source. For each underlined dream element, participants then described the waking event on a second form and indicated any similarities between the dream element and the waking event: Did they have the same people, objects, locations, or actions? Did they have the same themes and emotions? This affirmative probe was unusually complex, but the participants' reports conclusively demonstrated that our dreams are almost never an accurate replay of daytime events. There are, of course, exceptions, and we'll come back to some important ones in later chapters.

In some studies, we'll want to avoid using probes of any kind to ensure that our participants won't know what we're after. Or we might have no specific question other than what some group of participants' dreams look like. For example, we might want to know how the dreams of recently divorced women differ from the dreams of single women or of women who describe themselves as happily married. In such cases, we might not have any particular feature of dreams that we're especially interested in, so we just collect the most detailed dream reports we can.

Over the years, we've collected thousands of dream reports using many of the techniques described in this chapter. We love doing this. There's something magical about just sitting and reading a hundred reports. It can almost feel like studying dreams does them a disservice. But we want to know what dreams are like and how the brain creates them. And we want to know *why* the brain dreams. So let's turn our attention to these questions.

Chapter 5

SLEEP—
JUST A CURE
FOR SLEEPINESS?

WE ARE ALL CREATURES OF HABIT, CONTROLLED BY A collection of demanding animal drives. When the brain calculates that we should sleep, it lets us know. It might be the sandman making our eyes feel dry and itchy, or that feeling of heavy eyelids that we can't keep open. We might feel the desire to lie down, just for a second, or have a growing inability to pay attention and think coherently. The message from our brain comes through loud and clear: time for bed.

We know we must eat to get nutrition and drink to prevent dehydration. But the reason we need to sleep is something researchers are just beginning to understand. As recently as the late 1990s, we were largely in the dark about the biological function of sleep. There were arguments that sleep conserved energy, allowing organs, tissues, and cells to recover from the consequences of a day of exertion. Others argued that sleep just kept us out of harm's way at night. But strong scientific support for these ideas was lacking, and none of these arguments seemed strong enough to explain sleep's appearance and maintenance over hundreds of millions of years of evolution.

The suggestion by some researchers that sleep didn't serve any function led the pioneering sleep researcher Allan Rechtschaffen in 1979 to acerbically comment, "If sleep does not serve an absolutely vital function, then it is the biggest mistake the evolutionary process ever made." Even twenty years later, at the close of the twentieth century, things weren't much better: Allan Hobson at Harvard Medical School accurately quipped that the only known function of sleep was to cure sleepiness.

How important is sleep, really? To many people, sleep seems more of an inconvenience than something with inherent value. But let's be very clear about this. If you keep rats awake for extended periods, they *all die* within a month. In humans there's a hereditary brain disease, known as fatal familial insomnia, whose name says it all.

Trying to get by on little or no sleep has disastrous consequences. Close to eight thousand fatal car accidents in the United States every year are caused by people falling asleep while driving, and if you count all crashes that result in someone being hospitalized, the annual number is close to 50,000.[1] Moreover, lack of sleep (and the poor decision making that often comes with it) has been implicated in some of the greatest disasters of the twentieth century, including Chernobyl, Three Mile Island, and the *Challenger* explosion.[2]

Acknowledging that people might injure themselves and others through intentional sleep deprivation, the *Guinness Book of Records* will no longer consider attempts to break Randy Gardner's 1964 sleep deprivation record (11 days 25 minutes). Hospitals should also take heed. Medical residents report a 130 percent increase in accidents and 500 percent more near misses when driving home after shifts lasting more than 24 hours. And those who have more than five of these extended shifts in a month have both a sixfold increase in the number of medical errors they made and a threefold increase in fatigue-related patient deaths.[3]

At the end of the last century, there wasn't much hard data for sleep doing anything besides curing sleepiness—which is no small benefit. But over the last twenty years, a veritable explosion of scientific research has clearly demonstrated that sleep serves not one, but many critical functions. Although most of these functions aren't obviously relevant to our discussion of dreaming, having an appreciation of the full range of sleep functions will give us the background we need to understand why we dream.

Recent evidence suggests that some animals can go extended periods with only minimal amounts of sleep: mother whales with newborn calves, and birds migrating across large expanses of open water (some birds have also learned to sleep while in flight!). But there are no known animal species, including humans, that do not need to sleep. In fact, not one human has ever been identified who didn't need sleep. Even insects and roundworms have their own lowly versions of sleeping, becoming immobile and unresponsive for hours at the same time every day, even if kept in constant light or darkness. When these animals are deprived of their rest periods, they make up for them at the first opportunity.

For it to have been maintained across half a billion years of evolution, sleep must serve functions critical to our survival. Indeed, sleep serves some functions that by their very nature seem to require its presence. But other functions seem to have been almost casually assigned to sleep. These can be thought of as housekeeping functions, which have been assigned to sleep simply because it's a convenient time to get them done. Think about the cleaning of large office buildings every night. They're cleaned at night not because it's the only time when it could be done, but because it's more convenient for those who work in the buildings during the day and more efficient for the cleaning staff if no one else is there. But the cleaning could obviously be done in the daytime.

THE HOUSEKEEPING FUNCTIONS OF SLEEP

A number of items on the growing list of sleep functions fit into this housekeeping category. For example, consider how children grow.

During childhood this process is controlled by *growth hormone*, secreted by the pituitary gland located at the base of the brain. When too little is secreted, the child's height is stunted, resulting in a condition known as pituitary dwarfism. Conversely, too much secretion of growth hormone leads to pituitary gigantism.

During childhood, most growth hormone is secreted at night during deep, slow-wave sleep. The growth triggered by this release of growth hormone can cause a child to grow as much as two-thirds of an inch in 24 hours,[4] usually in conjunction with eating more and getting more sleep than usual. Indeed, in one study of infant growth, spurts often went hand in hand with bursts of increased sleep, and the infants grew up to a tenth of an inch for every extra hour of sleep.

There's no obvious reason why growth must occur during slow-wave sleep. Nothing about the changes in brain or body physiology during sleep seems to be crucial for growth spurts. Most likely, evolution pushed the time of release of growth hormone and its consequent growth spurt to slow-wave sleep simply because not many other demands were being made on the body at that time. It certainly makes more sense than timing growth to when a child is awake, running around and playing. Intuitively, it seems that growing taller must be easier to do when you're lying down, asleep.

The same can be said for sleep's other housekeeping functions, such as regulating insulin and producing antibodies. In the ten days after getting a flu shot, the blood levels of antibodies against the flu virus can increase fifty-fold. But to get the optimal benefit from the vaccination, its recipient must get enough sleep. In one study, participants were allowed only 4 hours of sleep a night for six nights, starting four nights before the vaccination. A week later, their antibody levels were only half those of participants who had slept normally.[5] In another study, participants deprived of just a single night of sleep after receiving a hepatitis vaccination similarly produced only half the antibody levels of controls.[6] Why is sleep so important for antibody production? Again, we lack a firm answer, but the reason is most likely the same as for growth hormone. Sleep is just the easiest time to get the immune response cranked up for maximum antibody production.

Insulin regulation tells a similar story. The hormone insulin is produced in the pancreas and secreted when levels of glucose in the blood start to rise. The hormone instructs muscle, liver, and fat cells to absorb the excess glucose from the blood and convert it to glycogen or, for longer-term storage, into fat. When this process breaks down, the result is diabetes. What does sleep have to do with this? After being allowed only 4 hours of sleep for five nights, otherwise healthy college students began to look prediabetic.[7] The rate at which their bodies cleared glucose from their blood dropped 40 percent compared to controls getting 8 hours of sleep a night, reaching levels seen in older adults with impaired glucose tolerance. In addition, their body's response to the acute administration of insulin dropped 30 percent, similar to the change seen with aging or in pregnancy-related diabetes.

Eve van Cauter at the University of Chicago has suggested that the current epidemic of obesity—40 percent of American adults are now considered obese, up from just 14 percent fifty years ago—may be due as much to our ever-decreasing sleep as to our ever-increasing intake of sweets. Again, we're not sure why sufficient sleep is critical for maintaining insulin's ability to effectively regulate blood sugar levels, but critical it is, and the trend toward less and less sleep may be leading up to 100 million Americans on the path to diabetes.

One last housekeeping function of sleep is worth noting. Sleep appears to play an important role in the cleansing of unwanted waste products from the brain, including β-amyloid, a protein whose accumulation in the spaces between nerve cells is a major determinant in the development of Alzheimer's disease. We don't know why β-amyloid accumulates in our brains as we age, but it's clear that it is removed from our brain twice as fast while we're sleeping as when we're awake.[8] After just a single night of sleep deprivation, the level of β-amyloid in these interstitial spaces increases by 5 percent.

In this case, we think we know why sleep is so important. The clearance of waste from the brain is linked to the flow of cerebral spinal fluid, which washes over the cells in the brain and carries away waste products. It turns out that during sleep this flow occurs in pulses, and the pulses are timed to the slow waves of nonREM sleep. It's these slow waves that appear to be driving the flow of cerebral spinal fluid and the clearance of β-amyloid from the brain.[9]

THE CRITICAL FUNCTIONS OF SLEEP

Although we classify all of these functions of sleep as "housekeeping" functions, there's nothing shabby about the list. They help us grow and keep us from getting sick, overweight, and cognitively impaired. But they can't explain why sleep evolved in the first place. Indeed, once sleep had arisen, it makes sense that these housekeeping functions migrated to the time of day when we are asleep. But there must also have been critical functions of sleep that couldn't be met during wakefulness.

From an evolutionary perspective, some of the best evidence that such a critical function exists may come from dolphins, whales, and some species of birds. Dolphins have a serious problem: If they were to fall asleep, they would stop swimming, sink, and drown. They can't afford to sleep. If sleep had housekeeping functions only, this wouldn't be a problem. Evolution could relatively easily produce a dolphin that simply didn't sleep. It would have to move all of those housekeeping functions into the wake state, but this would involve only relaxing the constraints on when the functions were performed. But instead, evolution found a much more difficult—almost unimaginable—solution to the problem. Dolphins and whales evolved the ability to have only half of their brain sleep at a time, switching from one side of the brain to the other every hour or so. Similarly, the great frigatebird (*Fregata minor*) will fly over the ocean for months without landing, resorting to "unihemispheric sleep" while in flight. In ducks, unihemispheric sleep is used to guard against predators. When a flock of ducks are sleeping in a pond, those around the outer edge keep the half of their brain that looks outward awake, switching positions with those in the center of the flock as the night goes on.

Such an evolutionary tour de force tells us there must be an absolutely unavoidable requirement for sleep. It can't be a function that simply requires lying down, closing the eyes, or being relaxed. Humans can do all these things without being asleep. Instead, these critical functions must require us to be cut off from the outside world, unaware of what is happening around us, and truly asleep. Offline memory processing precisely fits the bill. Our brains are not like DVRs, which can record an ongoing TV show while playing back an older recording to the screen. We cannot simultaneously pay attention to new sensory information and replay or analyze previously stored memories. It has to be one or the other. This happens all the time in conversations: Our eyes wander off to the side as we think about something else, or even as we think about what someone just said—it's one of those embarrassing moments when we have to ask someone to repeat themselves. That's why we need to sleep. We spend the waking day attending to our surroundings, taking in new information and storing it away, waiting until we sleep to review and revise this information and figure out what it means.

For every two hours we spend awake, storing away new information, our brain needs an hour of sleep to figure out the meaning and significance of that new information—an hour disconnected from the outside world and with the normal top-down mechanisms that guide our waking thoughts and actions turned off. That's the crucial task evolution has for sleep.

Sleep and Memory Evolution

When first encoded in the brain, memories are fragile, susceptible to both interference from other newly forming memories and simple forgetting. If not forgotten within a few seconds—as your memory of the start of this sentence will be—it remains in a relatively delicate form for a few hours, until the brain has had a chance to "consolidate" it. This process involves the synthesis of new proteins that cement the connections within a network of nerve cells that collectively form the physiological basis of the memory.

Memory consolidation was first described in 1900 by the German psychologist Georg Elias Müller and his student Alfons Pilzecker. But evidence that sleep played an additional and sometimes crucial role in this process came only much later, thanks largely to the work of Elizabeth Hennevin in France and Carlyle Smith in Canada. Together they published two dozen articles on sleep and memory in the 1970s and 1980s. Despite their consistently high-quality publications, it wasn't until 2001—a full hundred years after that first description of memory consolidation by Müller and Pilzecker—that an article from Bob's laboratory entitled "Sleep, Learning, and Dreams: Offline Memory Reprocessing" was published in the prestigious journal *Science* and finally pushed the research community into taking the idea of sleep-dependent memory consolidation seriously. Announcing a new era of sleep and dream research, it boldly proclaimed:

> *Converging evidence and new research methodologies from across the neurosciences permit the neuroscientific study of the role of sleep in off-line memory reprocessing, as well as the nature and function of dreaming. Evidence supports a role for sleep in the consolidation of an array of learning and memory tasks. In addition, new methodologies allow the experimental manipulation of dream content at sleep onset, permitting an objective and scientific study of this dream formation and a renewed*

search for the possible functions of dreaming and the biological processes subserving it.[10]

Since 2001, over a thousand scientific papers have appeared, extending our knowledge of how sleep stabilizes, enhances, integrates, analyzes, and even alters our memories, processes that vastly improve what we know and how we understand it. You may have noticed that we titled this section "Sleep and Memory Evolution," not "Sleep and Memory Consolidation." Although sleep does consolidate recently formed memories and make them more resistant to interference and forgetting, it also does much, much more than that. The term *memory evolution* acknowledges this, as well as the fact that our memories continue to change in a number of ways over our entire lifetime.

Pianos and Typing

Sleep can enhance many forms of memory, for example, the motor (muscle) skills gained by people who study a musical instrument or engage in sports such as gymnastics that involve learning complex movement sequences. It's not uncommon for a student practicing a Chopin étude on the piano to report getting stuck on some short passage that they can't master and giving up in frustration, only to come back the next morning and play it perfectly the very first time. When asked how they understand this phenomenon, these people usually say that, probably, by the time they had stopped practicing the day before, they had mastered it but were just too tired to play it. But they are wrong. It appears that they actually perfected it while they were sleeping that night.

We haven't studied pianists, but we have studied regular people learning to type the number sequence 4–1–3–2–4 on their computer. After 5 or 6 minutes of practice on this finger-tapping task, they've usually gotten about 60 percent faster, but then they plateau and by the end of the 10-minute training session haven't gotten any faster. We then send them away for 12 hours, after which we test them for just 1 minute. When we train participants in the morning and test them that evening, we find that they haven't forgotten what they learned; what we observe is them typing as rapidly as at the end of training, although no better. But when we train them in the evening and bring them in the next morning after a night of sleep, they are 15 to 20 percent faster and are making fewer errors. Overnight, their sleeping brain actually improved their ability to type the sequence. Similar sleep-dependent improvements are seen for learning visual and auditory discrimination skills. In all these cases, participants show dramatic overnight improvements while showing no improvement over equivalent periods of daytime wakefulness. As Matt Walker at Berkeley summed it up, "practice *with sleep* makes perfect."[11]

Word Play

In some cases, sleep actually makes your memory worse, although probably more useful. In one study that showed this effect, Jessica Payne, who was a postdoctoral fellow in Bob's lab at the time, had participants listen to several lists of words, telling them to try to memorize the words. For example, one of the lists included the words *nurse, sick, lawyer, medicine, health, hospital, dentist, physician, ill, patient, office*, and *stethoscope*. Then, either just 20 minutes later or 12 hours later, she asked them to write down all the words they could remember. In other studies, participants were tested by giving them a new list of words and asking them which words on the new list were also on the old list.

You can try this yourself. Without going back and looking at the list above, try to remember which of these words were on that list: *cotton, medicine, patient, table, doctor, letter*. If you picked *doctor* as one of the words from the original list, you're not alone, although you are wrong. As many as half the participants in Jessica's study wrote that they remembered hearing *doctor*,[12] which wasn't that surprising. They made this mistake because the original list was made up of words that most often come to mind when given the word *doctor*. Because all of the words in the original list are so strongly associated with *doctor*, your brain accurately calculates that the word *doctor* represents the gist of the list and then erroneously concludes that it must have been on it. Thus, *doctor* can be thought of as the title of the list. The other lists were constructed the same way, around other words that, in every case, were not included in the list.

What happened in this experiment? Whether participants spent the 12 hours before testing awake or asleep, they forgot 30 to 40 percent of the words they had initially memorized (The initial level was measured in the groups tested after just 20 minutes.) But for the title words that they erroneously remembered hearing, participants who spent a day awake before testing "forgot" about 20 percent of them while those who got a night of sleep before testing actually "remembered" 5 to 10 percent *more* title words in the morning. Sleep selectively stabilized and perhaps even enhanced the false memory of these title words while forgetting the actual words on the lists. Interestingly, this phenomenon is not unlike a dream.

Dreams don't replay memories exactly; they create a narrative that has the same gist as some recent memory and could have the same title. This is our first example of how memory evolution during sleep is similar to dreaming, and we'll come back to it when we talk about the function of dreams in Chapter 7. But for now, notice that this is an example of dream content matching memory processing in sleep.

It's true that in this case, sleep technically worsened participants' memory—they remembered more of the gist words that weren't actually on the list. But we believe that's the wrong way to think about it. Unless you're taking a test in school or testifying in court, remembering things perfectly is rarely the point, and it's unlikely that our memory systems evolved with a goal of "total recall." Instead, evolution aimed for a system that remembered what would be most useful in the future.

Dan Schacter, a professor of psychology at Harvard, has argued that memory is about the future, not the past.[13] Memory didn't evolve to give us something to reminisce about in our old age; memory evolved because those who don't learn from the past are doomed to repeat it. Memory evolved to give us a leg up when we run into a situation similar to what we've encountered before. Thus, when faced with a list of related words, an efficient memory processing system may preferentially extract and retain the general theme (or gist) of the list rather than focusing on the specific words. Indeed, how many words you can recall from the lists depends on how many gist words you remember, because they help you remember the lists. Initially the brain seems to memorize both the gist and the actual words, but there's a limit to how effectively it can store a hundred words in long-term memory. It appears that the brain needs time off from other tasks—it needs to sleep—to decide which is more important.

Defining Who We Are

Sleep also plays a major role in forming our sense of self. How we think about who we are is determined largely by our autobiographical memories of important events in our life, and sleep helps shape these memories. Several laboratories have shown that sleep preferentially consolidates emotional memory, leaving less-interesting memories to be forgotten. Jessica Payne extended these findings by showing that sleep even selectively consolidates the emotional portion from a photograph of a scene (for example, a crashed car but not the palm trees in the background), leaving the rest of the details in the photo to be forgotten.[14]

Just the emotional objects in the photos benefited from a night of sleep—not any unemotional objects, not the neutral backgrounds—and it is these same key emotional elements in our autobiographical past that we most remember and use, both consciously and unconsciously, to construct our sense of who we are. In a real sense, we are what we sleep. And, of course, our dreams likewise capture the emotions of waking events much more than the details of those events, making this a second example of dream content matching memory processing in sleep.

Sleep can also soften our emotional responses when we recall them. Matt Walker, who conducted the finger-tapping typing experiment with Bob, has coined the phrase, "Sleep to forget, sleep to remember"[15] to describe this process. Although sleep selectively retains our emotional memories, it reduces the strength of our emotional response when we are exposed to them again. Such softening of our emotional responses is a crucial element of recovering from traumatic events—and, again, we can thank sleep for this benefit.

Understanding the World

Sleep can also discover patterns in our daily events, finding rules for how the world works that aren't accessible to our waking brains. One study demonstrating this amazing ability of sleep was carried out by another of Bob's postdocs, Ina Djonlagic, a neurologist specializing in sleep disorders at Harvard Medical School.[16] The learning task Ina used is called the weather prediction task. It's a little like the game of blackjack, but it uses a deck of cards containing just the four aces (Figure 5.1). For each "hand" of the game, a participant is dealt one, two, or three of the aces and then must predict whether the dealer is holding a "sun" card or a "rain" card.

At first participants have no way of even guessing what, for example, a hand consisting of the ace of hearts and ace of clubs might tell them about the dealer's card. But in each of 200 training trials, they are told what the dealer holds, and so they gradually get a sense of how their hand might predict what the dealer is holding. What makes it tricky is that it's a probabilistic task; sometimes a given hand will predict rain, but other times it predicts sun, and the four aces individually predict rain at rates ranging from a quarter of the time to three-quarters of the time. Still, when subsequently given 100 test tri-

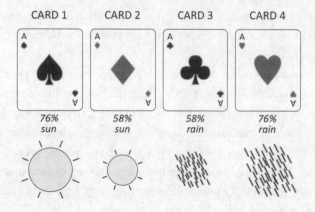

Figure 5.1. Prediction probabilities for cards
in the weather prediction task.

als, most participants managed to get 70 to 80 percent of them right, correctly picking the card most likely to be in the dealer's hand. The odds of someone getting that many by chance are less than one in a million, so we know the participants are learning the task reasonably well. Still, they're never perfect. They develop a sense of how it works, but they don't actually figure out the rules.

Can sleep help them improve? Participants who were trained and initially tested in the morning and then retested that evening still remembered what they had learned in the morning, which is pretty impressive. But they weren't noticeably better than they had been in the morning. In contrast, when those trained in the evening were retested the next morning, they did 10 to 15 percent better.

Somehow, participants in Ina's study had a better understanding of the test the next morning. And though it's only a small part of their world, in a very real sense they understood how the world worked better in the morning than they had when they went to bed the night before. In fact, we believe that for each of us, much of our understanding of the world is constructed from thousands of nights like this one. In some cases, remembering a dream can have the same effect. As another example of how dream content matches sleep-dependent memory processing, more than one Nobel laureate has attributed their prize-winning discoveries to dreams that revealed to them how some piece of the world worked.

Even infants use their sleep to extract patterns from the world around them. Rebecca Gomez at the University of Arizona in Tucson studies sleep and memory in infants and has shown that infants can learn an artificial grammar—a set of rules explaining how invented words are constructed—amazingly quickly, but they need a nap after learning if they want to remember it.[17]

Here's how Gomez showed this. She created 48 three-part nonsense words like *pel-wadim-jic* and *vot-puser-rud*. Half the words began with *pel* and half with *vot*, and the grammatical rule was that words starting with *pel* always ended with *jic*, while words starting with *vot* always ended with *rud*. A recording of words was played over and over for 15 minutes while the infants played quietly.

Four hours later, the children who had napped for at least 30 minutes after hearing the recording showed that they knew the grammar. They responded with surprise whenever they heard an ungrammatical word—for example, one starting with *pel* but ending with *rud*. But they did not seem surprised when they heard a word that met the grammatical patterns—such as one starting with *jic* and ending with *rud*—even if they had never heard the word before. The next morning the children who had gotten a nap during those first 4 hours still remembered the grammar; those who didn't nap appeared to have forgotten it.

One of the reasons infants need to nap appears to be that, unlike adults, infants can't maintain new memories long enough to wait all day before they sleep on them. This might even be the reason they get so crabby when they don't get their nap. Infants are nonstop learning machines, and without periodic naps to allow their sleeping brains to process the new information their little brains are taking in, they get overloaded. They start feeling burned out, the same way adults do who have taken in too much information without a break.

Creativity and Insight

Perhaps the most impressive form of sleep-dependent—particularly REM sleep-dependent—memory evolution is the ability to enhance creativity and insight the next day, benefits that are also seen after dream recall. In a study conducted by Sara Mednick using the Remote Associates Task, a nap was enough to help participants figure out the single word that linked three others together. Perhaps you can figure out the missing link in Figure 5.2.[18]

Don't worry if you can't, but you may want to try again after you get some sleep. REM sleep provides a brain state in which weak and unexpected associations are more strongly activated than normally strong associations,[19] explaining how it aids in finding the remote associates and perhaps explaining the bizarreness in our REM sleep dreams.

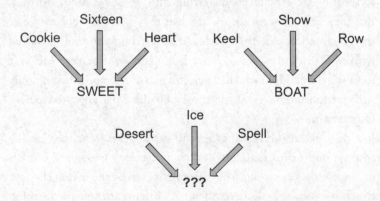

Figure 5.2. Examples of the Remote Associates Task.

Problem Solving

Somewhere between, on the one hand, the kind of pattern recognition seen in the weather prediction task and the grammar learning of infants and, on the other hand, the creativity and insight of the missing link task just described, is plain old problem solving. Whether you sleep on a problem (the English version) or take a problem to bed with you (the French version), we all have an intuitive knowledge that sleep helps us make difficult choices.

Faced with the choice of taking a boring but high-paying job or an exciting but much lower-paying job, what do we do? We "sleep on it." More often than not, when we awaken the next morning, we find the decision made. There's no assessment of the alternatives provided—no explanation of the rationale for the choice—just the choice. As in the weather prediction task, we know our choice without being able to explicitly describe the basis for the decision—still, it seems we can usually count on it to be the right choice.

Why We Have Different Stages of Sleep

Sleep provides a unique benefit for all of these different forms of memory evolution. But the different stages of sleep don't contribute equally. For example, overnight improvement on the typing task depends on how much N2 sleep we get, especially late in the night. Most verbal memory tasks depend on how much N3 sleep we get, while emotional memory and problem solving tasks seem to depend on REM sleep. A visual discrimination task's overnight improvement depends both on how much N3 sleep you get early in the night and how much REM sleep you get late in the night.

These unique sleep-stage dependencies may explain why all these different sleep stages evolved in the first place. If we think of sleep as a time when the brain is optimized for memory evolution, it makes sense that the neurophysiology and neurochemistry that would be ideal for strengthening memories for word lists might be different from what's ideal for improving on a typing task, and both might be different from the best conditions for problem solving. As far as we know, this is the best explanation available for why humans have so many different stages of sleep.

WHEN THINGS DON'T WORK RIGHT

In science, the exceptions often tell us more about the rules than any example of the rules working. An example of this for our investigation of sleep and memory is post-traumatic stress disorder (PTSD). Following a traumatic event, our brain forms a raw, detailed, often overwhelmingly emotional memory (Figure 5.3).

In most cases, these memories are processed by the brain, largely without intent and outside of awareness, leading to the resolution of the trauma. The trauma is not forgotten, but its rough edges are worn off. The memory is no longer intrusive, popping into the person's mind whenever anything remotely similar is encountered. And when it is recalled, the details are forgotten, the emotions that arise are less intense, and there is some sense of how to understand the traumatic

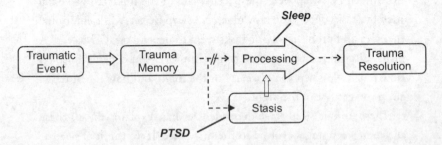

Figure 5.3. PTSD as a consequence of impaired sleep-dependent memory evolution.

event so that the person can move forward with life. In cases when this processing doesn't occur, the memory becomes locked in stasis, a condition known as PTSD. When we look at the changes that failed to occur—the loss of detail, the weakening of emotional response, the evolution of understanding—we see that these are all processes that normally occur best, and maybe only, during sleep. From this perspective, we can think of PTSD as a disorder of sleep-dependent memory evolution.

We'll have more to say about this disorder in Chapters 7 and 13 when we discuss the function of dreaming, because a hallmark of PTSD is the occurrence of nightmares that are near-perfect replays of the traumatic event. Such near-veridical replay of our waking events in dreams is something that normally never happens, and knowing that it does in PTSD will give us insight into the function of dreaming and how this function breaks down in those suffering from PTSD.

BEFORE WE COULD DISCUSS the functions of dreaming in Chapters 7 and 8, we needed to understand the biological functions of sleep, which are more easily defined and measured than those of dreaming. Sleep serves many such functions. But in the end, the role of sleep in emotional and memory processing is most tightly tied to dreaming, and the contributions of sleep to the evolution of memory presages our discussion of the function of dreaming. We repeatedly saw the links between this memory processing and the characteristics of dreams. Studying why we dream, however, is a very different proposition from asking why we sleep. The reason for this difference is the uniquely subjective nature of dreaming. So, before we can address the question of why we dream, we need to take a brief side trip into the conundrum of consciousness. Then we'll be able to look at how it all fits together.

Chapter 6

DO DOGS DREAM?

HERE'S A QUICK TEST TO START OFF THIS CHAPTER: Which of these groups dream: adults, babies, dogs, people in comas, rats, books? Now rate your confidence in each of your answers. We suspect that you'll be absolutely convinced that adults dream and that books don't. But we suspect that you'll be somewhat less sure about babies and dogs, although you probably believe that they do. And frankly, we have no idea what you'll say about rats and people in comas. Each of these groups raises interesting questions about dreaming, so let's look at them in more detail.

❖ ❖ ❖

IF YOU ASK ANY DOG OWNER whether their dog dreams, odds are they'll tell you that it definitely does. But how do they know? Well, they might say that their dog barks quietly or whimpers while it's asleep, and its legs start to twitch as if it were running. Obviously, it's dreaming about chasing or being chased by some other animal. You can Google "dog dreaming" and find a delightful—or maybe frightening—collection of videos of this behavior. It's hard to believe that these dogs are *not* dreaming. You can read an article posted on the *Psychology Today* website and find out that "It is really quite easy to determine when your dog is dreaming. . . . All that you have to do is to watch him."[1]

Indeed, dogs do spend a considerable amount of time in REM sleep, just like us. They even appear to spend more time—36 percent compared to our 20–25 percent.[2] And they definitely do twitch, whimper, and display running-like leg movements while they're sleeping. So yes, it certainly *looks* like they're dreaming.

Similarly, perhaps even rats dream. MIT neuroscientist Matt Wilson records electrical activity from neurons deep in the brains of rats while they're in REM sleep.[3] While the rat is in "dream sleep," he told the *New York Times*, it's "certainly recalling memories of [running around a track] as they occurred [earlier] during the awake state."[4] But he stopped short of saying they were dreaming, because the dreaming experience is, as we know, subjective. As Wilson drolly commented, "Our ability to ask the animal to report the content of these states is limited." Ah, there's the rub. Like dogs, the rats *appear* to be dreaming. But, as we saw in Chapter 1, making sense of the internal, private, and imaginary aspects of dreams is tricky business, and it is only with considerable effort that humans come to understand the nature of their own dream experiences.

If rats experience consciousness when they're awake, it's reasonable to consider whether they dream. But if they do indeed dream, this finding would suggest that they experience a subjective awareness of running around a track or of being rewarded with food while in their dreams. But how does this sphere of awareness relate to their conscious experiences during wakefulness? Do rats remember their dreams and, if so, can they distinguish dreamed events (for example, having just eaten or navigating a maze) from reality (waking up hungry and inside a cage)? And how do they *understand* dreaming? Given the difficulty children have coming to understand that dreams are, well, *dreams*, it's hard to imagine that rats could comprehend that remembered dreams don't reflect reality. Frankly, we doubt that they would remember them at all.

As you've probably noticed, we've wandered into a thorny problem—one much larger than whether dogs dream. It's the question of whether animals other than humans are even conscious. Bob likes to shock his students by telling them that there is no scientific evidence that humans are conscious. It's true. As the philosopher David Chalmers noted, "There is nothing that we know more intimately than conscious experience, but there is nothing that is harder to explain."[5] It's a subjective, internal sense of self—we experience things in our mind, we feel our emotions. To paraphrase the French philosopher René Descartes: We think, therefore we're conscious.

All scientific studies of consciousness—or rather, all studies of consciousness, scientific or otherwise—start by assuming that humans are conscious. Only then can we ask questions about this state, such as which regions of the brain are active when we consciously perceive a visual stimulus or whether we're conscious when we're in deep sleep. But scientists need to start with the assumption that we do *experience* consciousness, and, usually, that we're conscious when we say we are, which makes sense. We can make that assumption safely, but we all know this only because each of us experiences our own consciousness. We are aware that we are, in fact, aware, but no one has ever proven it scientifically.

It's equally safe to assume that the book in your hands is not conscious. If we asked you what you thought it felt like to be a book, you'd probably answer, "Nothing." The question doesn't even seem to make sense. But it's quite a good question. The philosopher Thomas Nagel first asked it back in 1974;[6] not about a book, but about a bat (the animal). He wasn't really interested in *what* it was like to be a bat; more fundamentally and importantly, he wanted to know whether there was *anything* it was like. For there to be an answer to the question, he argued, the bat must be conscious; it must have some kind of personal experience, some feelings, perceptions, or thoughts that *it is aware of*—something that could form the basis for saying what it's like to be a bat. That, he argued, is the test for whether something is conscious. When you say to yourself, "I wonder what it's like to be

Fido," you're starting from the assumption that there is *something* it's like to be Fido.

But as things currently stand, no matter how much evidence you might have, you can really only say that Fido *acts* like he's conscious, which is not enough. Within twenty years, you'll probably be able to buy *eFido*—sold by some tech company—that acts just like Fido. Maybe its movements are a little jerky, and true, it doesn't drool or poop, but it jumps up and down when you pick up its leash and lowers its head ashamedly when you catch it lying on the pillow it took off your bed. And for all intents and purposes, he looks as if he *loves* you. We're so convinced of his humanity that in the last sentence we switched from talking about "it" to talking about "him." But eFido isn't conscious. He's just an elegantly programmed robot. How do we know? Well, to be honest, we don't know for sure, because we simply don't know what it takes for something to be conscious. But we do hope he isn't.

On the other hand, you can be absolutely certain that the snowman character called Olaf in the movie *Frozen* isn't conscious, and neither is the heroine Elsa. They are, after all, computer-generated images. We bring up this example not because we think you'll disagree, but to point out that many five-year-olds would rather give up their belief in Santa Claus than accept that Elsa isn't a living, breathing, conscious child, just like them. We attribute consciousness to anything that acts like us. It's that simple.

And it's also that confusing. That's the problem with asking whether dogs dream. They certainly *act* like they're dreaming, and their brain activity while they're sleeping looks very similar to ours when we dream. But neither of these facts allows us to conclude that they're dreaming. We just can't tell. And that's not the worst of it.

✧   ✧   ✧   ✧

WE LOVE TO WATCH little babies when they're dreaming. They smile, their legs jerk and their arms wave around, they make little noises, and they start to suck on imaginary breasts, all while they're in REM sleep. You can tell that they're in REM by watching their eyelids. Their lids are still so thin, you can clearly see the lids stretch out as the eyeballs leap back and forth—right, left, right, left. Babies also get into REM sleep much faster than we do—after being asleep for less than 10 minutes[7] compared to 90 minutes for adults—so you don't even have to wait very long to see it start. It's really amazing to watch—the moving eyes, the smiles, the sucking noises. How could they not be dreaming?

You can see where this is going; we have the same problem as with the dogs and rats. Yes, babies *act* like they're dreaming, but we just can't know. They can't tell us their dreams; they can't even tell us that they're conscious. And if they are conscious, we clearly don't have a good grasp on the many ways in which their sense of consciousness differs from that of older children and adults, or how it changes as the brain grows and develops over the first few years. If they are conscious and if they do dream, it would certainly be with a diminished, preverbal form of consciousness and dreaming that would become more like ours only as they grew older. What they could dream about would be limited by the knowledge and skills they had already gained.

In the first weeks of life, a baby's vision is pretty blurry beyond a distance of about 12 inches, and they can't turn their eyes to follow a moving object until they're a few months old. Thus, their visual memories are going to be quite limited. Similarly, their sense of self would be minimal, perhaps just a sense of being. They certainly won't have a vault filled with rich autobiographical memories the way adults have, and they won't be able to categorize their sensory experiences. Without this type of information, babies' dreams would be little more than a pastiche of remembered sensations and emotions.

Of course, as babies grow older and as their experiences, memories, and cognitive skills expand, so too will the richness of their dreams. This is exactly what laboratory and home studies of children's dreams suggest: As children's brains mature and their cognitive abilities improve (including the complexity of their mental imagery and narrative skills), the complexity of their dream experiences and their capacity for dream recall also show developmental changes. This growth results in recalled dreams that are increasingly adult-like in frequency, length, and structure.[8]

So, assuming they are conscious and do dream, what *do* babies dream about? No one knows for sure, but we believe that their dreams are probably similar to whatever their *waking* experiences are like, and certainly not more complex. Newborns aren't going to be dreaming about doing puzzles, driving a car, or looking for a lost dog. What they're dreaming is probably just what it looks like: images of happy faces, the warmth of being held close, and the sensations of sucking on bottles and breasts.

❖   ❖   ❖   ❖

HOW ABOUT ADULTS? Do all adults dream? Compared to dogs and babies, this question is an easy one. We all dream, right? Well, maybe. The vast majority of adults—at least 85 to 90 percent—will tell you that they dream. In addition, studies have shown that most people who say they never dream actually do. When brought into the sleep lab and awakened from REM sleep, most of these people report that they were—to their amazement—dreaming.[9] Like the rest of us, they were dreaming all along, but just didn't remember their dreams after waking up.

Well, that's true for most people who say they don't dream. Jim Pagel at the University of Colorado has been studying these "non-dreamers" for years, and over a five-year period he found 16 patients in his clinical sleep laboratory—about one out of every two hundred patients—who reported never having dreamt. When he woke each patient once from REM sleep and once from nonREM sleep, none of them reported that they had been dreaming. So perhaps a half percent of adults actually can't remember any dreams, and it is possible these people never dream.

Of course, as you might expect, people who have suffered some kinds of brain damage appear not to dream. In a landmark study of stroke victims, South African psychoanalyst and neuropsychologist Mark Solms described a number of patients who reported a total loss of dreaming after sustaining damage to brain structures deep in the front of the brain.[10] But like the half percent of patients who came to Jim Pagel's clinical sleep laboratory, those patients are clearly the exceptions to the rule. Just about everyone dreams. And even for the few who appear not to, we can't be sure whether they're not dreaming or simply unable to remember their dreams.

LET'S PUSH THE ENVELOPE even further, and ask about people in comas. Do they dream? When we think about people in coma, we're right back where we were with infants. Because they can't report what they're experiencing, we can't tell, either. Here too, the question of their consciousness is a more primary one, and although scientists agree that patients in deep comas are unconscious, for patients in lighter forms of coma the answer is less clear. For example, patients in what is called a *persistent vegetative state* exhibit periods of apparent wakefulness—their eyes are open and sometimes seem to look around. But they don't respond to people and show no signs of self-awareness or awareness of their environment; these people are presumed to be unconscious. On the other hand, patients in a *minimally conscious state* may have some residual consciousness. Although very similar to those in a persistent vegetative state, patients in a minimally conscious state can look awake and display some limited signs of purposeful behavior, such as following an object with their eyes as it moves in front of them or even moving a finger when asked to. As the name of this condition suggests, scientists think that such patients are conscious to some degree, even though they can't communicate with others or even maintain attention for more than a second or two. What it's like to be minimally conscious is still unknown.

Do these patients dream? When researchers in Belgium, Italy, and the United States combined forces to study the sleep of patients in vegetative and minimally conscious states, they found that while people in a vegetative state showed no EEG evidence of normal sleep, those in a minimally conscious state did.[11] At night, five of the six minimally conscious patients looked like they were asleep: Their eyes were closed, and they weren't moving. They also showed signs of normal sleep stages, both nonREM and REM, and their REM sleep came mostly at the end of the night as it normally does. As far as the researchers could tell, the sleep of minimally conscious patients is indistinguishable from that of fully conscious people.

But were these patients dreaming? At least one author of the study, Steven Laureys, appears to think so. "Everything... indicates that they have access to dreaming," he told reporters at *ScienceDaily*.[12] Still, we only know that these patients, like dogs and babies, show brain activity consistent with dreaming; they look like they are dreaming. Whether they actually are remains a mystery.

❖   ❖   ❖   ❖

YOU MIGHT THINK that by now scientists would be able to tell when someone is dreaming without having to wake them up and ask. Unfortunately, we can't, although it's not a problem limited to dreaming. It's the whole consciousness problem again. We simply can't measure internal thoughts and feelings. They're not that different from pain. We never observe pain directly. Instead, we infer its presence through people's behaviors, such as grimacing, and through subjective reports ("I feel a burning pain here, and a shooting pain over here"). But these inferences are uncertain. A grimace might be a reflection of physical pain, but it could also reflect a sudden realization that it's raining and you have to take the garbage out. In fact, the inference can be stronger for pain than for dreaming, because when it comes to pain, we can look at activity in brain regions that are strongly associated with pain perception. In contrast, there simply aren't any known brain regions whose activity during sleep reliably predicts dream recall.

It's also difficult to infer that someone is dreaming based on their behavior. Bob's wife, Debbie, sometimes laughs in her sleep; if he wakes her up, she invariably tells him that she was dreaming. In these instances, Bob can reliably infer that she's dreaming without having to wake her up. But such inferences aren't always reliable. In a now-classic laboratory sleep study carried out fifty years ago by Alan Arkin at City College of New York, 28 chronic sleep-talkers were repeatedly awakened while sleep-talking and asked what they were dreaming. To everyone's surprise, what they said while sleep-talking matched what they subsequently said they had been dreaming about less than half the time.

In one condition, however, sleep behavior and dream content are almost always in agreement. REM sleep behavior disorder (RBD) is a medical condition in which the paralysis that normally accompanies REM sleep breaks down. We'll have more to say about this in Chapter 13 when we talk about disorders of dreaming, but the short version is that people with RBD often act out their dreams. Laboratory studies have shown that when this happens, the dreams reported by these patients almost always match the actual behaviors observed. So here, it seems we can be sure that someone is dreaming without having to wake them up.

Except, perhaps we can't. While dreams recalled after these episodes almost always match the observed behaviors, there are people with RBD who insist they never dream![13] In one case, a seventy-two-year-old Frenchman being recorded in the sleep lab started to kick with both legs, turned his head, and yelled (in French), "That's something, huh? Come on, say it!" Then he sat up, mumbled, and went on, "You took something from me, you'll get it in the mouth.

No slippers? What is it?" Then he actually stood up, threw things off his bedside table and boxed the wall, shouting: "You're going to get it in the mouth. . . . Do you want to get it in your mouth, there? Ouch!" All the while, his sleep recording showed that he was clearly in REM sleep. Yet on being awakened moments later, he had no memory of having been dreaming—a response supporting his claim that he never dreamed.

Does this sound familiar? Either he does dream, as his behavior would suggest, and he can't remember it when he awakens, *or* he's showing the behavior of someone who is dreaming but doesn't experience it as a dream. As with dogs, rats, babies, and people in comas, it *looks like* he's dreaming; but we simply can't know.

Even for people with good dream recall, we still can't tell when they're dreaming, although researchers are trying. Francesca Siclari, a neurologist at the University of Lausanne in Switzerland, led a study with Giulio Tononi at the University of Wisconsin that looked for the EEG signature of dreaming.[14] In the end, they found an area in the back of the brain where a combination of exceptionally low levels of slow waves and high levels of fast waves in the EEG reliably predicted that a participant was dreaming. It also showed the opposite— exceptionally high levels of slow waves and low levels of fast waves predicted that they weren't. But these combinations didn't occur often, and for the vast majority of the night, the study team couldn't predict whether people were dreaming or not. Still, give researchers enough time, maybe ten or twenty years, and we suspect they'll get pretty good at using brain imaging to tell us when someone is dreaming. Unfortunately, even with improved brain imaging, the results won't tell us whether people in comas or babies (or dogs or rats) are dreaming. We will only know whether their brains look like they're dreaming.

If you find yourself feeling confused at this point, you're not alone. At the start of this chapter, we quoted Matt Wilson warning a *New York Times* reporter that we can't know for sure whether rats dream.

But in an MIT press release about the same study, Wilson is quoted as saying, "We know that they are in fact dreaming and their dreams are connected to actual experiences."[15] Similarly, Dan Margoliash, who studies songbirds at the University of Chicago, has been quoted as saying, "From our data we suspect the songbird dreams of singing,"[16] even though it's obvious that no one knows "what it's like" to be a bird, let alone a dreaming bird.

We haven't managed to answer many questions in this chapter, but we hope that you can now appreciate the difficulties that sleep researchers face when talking about consciousness, and specifically about dreaming, from a scientific perspective. In the end, however, when we take off our scientist hats, we suspect that all of us would agree that most adults dream, that babies and dogs probably dream, and that people in a deep coma most likely don't dream. And we really don't have a clue about whether dreaming takes place among people in minimally conscious states, rats, whales, or songbirds. Hopefully we can all agree that the *Frozen* characters Olaf and Elsa don't dream.

Chapter 7

WHY
WE DREAM

WITHOUT A DOUBT, THE BIGGEST QUESTIONS ABOUT dreaming are all variants on this question: *Why* do we dream? This is, in fact, three separate questions: (1) How does the brain create a dream?, (2) What function does dreaming serve?, and (3) Why do we have to experience the dream for this function to be fulfilled? The short answers are: (1) Not sure, (2) Not sure, and (3) Not sure. But we've got some pretty good ideas about what the most likely answers are.

HOW DOES THE BRAIN
CREATE A DREAM?

Dreaming depends on the generation of patterns of brain activity that instantiate the content of the dream as it progresses over time. If I see my mother in my dream, then a neural representation of the visual appearance of my mother must first be activated in my brain. This isn't any more difficult for the brain than calling up her image right now, while I'm awake, and it may not be that different, either.

Indeed, we now have evidence that the patterns of brain activity generated when we see an object in the real world are pretty much the same as when we imagine the object, whether in waking or in a dream. The evidence comes from functional magnetic resonance imaging (fMRI) studies, which can record activity throughout the brain over several minutes or hours. With fMRI, the brain is divided into about 50,000 voxels—three-dimensional equivalents of your camera's two-dimensional pixels. Each voxel is a cube about a tenth of an inch on a side, and a snapshot of the activity in each of these voxels is taken every 2 or 3 seconds. Using an exciting new technique called multivoxel pattern analysis, researchers can determine the precise pattern of voxel activation produced in visual processing regions of the brain by a specific image, say, a picture of a baseball, or the average pattern of voxel activation for a category of images, for example, faces, tools, or doors. Then, using these *classifiers*, they can have a study participant look at pictures and reliably predict whether they're looking at a baseball, a face, or a door, entirely based on their pattern of brain activation.

This amazing tool has allowed us to confirm that the pattern of activity seen when you look at a face is also activated when you bring an image of that face back to mind from memory. But even more exciting is groundbreaking work led by Tomoyasu Horikawa, a young researcher at the ATR Computational Neuroscience Laboratories in Kyoto, Japan, who has now demonstrated that the same multivoxel patterns are activated when faces appear in our dreams.[1] In a study right out of a science fiction story, Horikawa and his colleagues calculated classifiers for several categories of visual images from the fMRI signals generated while participants viewed thousands of pictures. When they matched these classifiers to the participants' brain activity just before awakening them for dream reports, the research team found surprising agreements between the classifiers that best matched the dreaming brain's activity and the content of the dream reports. After one awakening, a participant reported, "Well, there were people, about three people, inside some sort of hall. There was a male, a female, and maybe like a child. Ah, it was like a boy, a girl, and a mother. I don't think there was any color." When the brain activation from the preceding 15 seconds was analyzed, the computer created a composite "best fit" classifier image that contained the same multi-voxel patterns of activity as when the subject had actually been looking at pictures of women and children.

From one point of view, this means that we can kick our problem down the road. The patterns of brain activity that represent the images in our dreams are created by reactivating the patterns originally produced when we saw similar images in our waking life. Reactivations of other patterns of waking brain activity undoubtedly generate the brain patterns of thoughts and emotions that we experience in our dreams as well.

"Wait a minute!" you might complain. "That doesn't explain how the narrative story that makes up the dream is created, and besides, I see lots of things in my dreams that I know I've never seen in real life." Both of these objections are legitimate. But again, we can turn these questions over to those who study the thoughts and images generated when people are daydreaming. If we asked you to imagine a bald baseball player with green skin and a potbelly, we doubt that you'd have much trouble doing this. And if we asked you to go on and imagine him swinging at a pitch with a wooden broom, you could probably do that, too. When brains dream, they almost undoubtedly create such images in exactly the same manner.

"But wait again!" we hear some of you saying. "You're avoiding the whole issue of *how* and *why* a particular dream is put together the way it is. My dreams never really replicate events from my waking life, and no one tells me to imagine them. So, where do they come from?" That's an excellent but considerably harder question. We'll need to put off part of it until later in this chapter, but we can answer part of it here.

When Allan Hobson wrote *The Dreaming Brain* some thirty years ago, he argued that trying to determine how the brain constructs the specific content of individual dreams was a fool's errand. (At the time, functional brain imaging didn't exist.) Instead, Hobson argued, researchers should seek to understand the formal properties of dreams. Why were dreams so bizarre? Why were they so emotional? Why were they so heavily dominated by visual imagery and movement? In *The Dreaming Brain*, Hobson argued that these phenomena were simple consequences of the neurophysiology of REM sleep. (Both in *The Dreaming Brain* and in his later writings, Hobson has had something of a love-hate relationship with nonREM dreaming, on the one hand acknowledging that it occurs, but on the other dismissing it either as the result of REM physiology creeping into nonREM sleep or as just plain uninteresting.) Hobson catalogued the unique physiology of REM sleep and compared it to these formal properties of dreaming.

Eight years later, Pierre Maquet and his colleagues in Belgium published the first of several functional brain imaging studies of REM sleep and dreaming.[2] The broad patterns of regional brain activity they found seemed capable of explaining many of the formal properties of dreaming described by Hobson. During REM sleep, brain activity increases in large portions of the limbic system, which mediates emotional expression. At the same time, activity *decreases* in a structure with the tortuous name of dorsolateral prefrontal cortex; this area of the brain plays a critical role in executive functioning—in planning, logical reasoning, and impulse control. These neurophysiological changes, the research team argued, are sufficient to explain why most of our dreams contain emotions as well as why our dreams seem to lack judgment, planning, and logical reasoning.

At a finer level of description, the preferential activation of weak associations mentioned in Chapter 5 provides an explanation for the bizarreness of our dreams. Similarly, in REM sleep the activation of the motor cortex, a brain region that normally controls our movements, undoubtedly produces the sensation of movement within our dreams.

Together, these findings make us reasonably confident that we know how the brain generates the patterns of activity necessary to form the internal representations of the images and movements of our dreams, as well as why it generates their often bizarre and emotional flavor with a seeming lack of judgment, planning, and reasoning. But we still haven't explained how or why specific dreams are created.

WHAT FUNCTION DOES DREAMING SERVE?—TAKE 1

Why do we dream? What evolutionary need does dreaming fulfill? What does dreaming *do*, beyond plunging us into an often strange, all too vivid and convincing dream world? Whatever the function of dreams may be, it simply cannot depend on our remembering them once we awaken. As we saw in Chapter 4, people dream in all sleep stages; so we're immersed in various forms of dreaming for at least two-thirds of the night—more than six hours out of eight—and some researchers would even say we dream all night long. If you're one of those lucky souls who falls asleep quickly and sleeps soundly through the night, it's unlikely that you recall even 5 percent—20 minutes—of those dreams. Moreover, even if we limit ourselves to only the most vivid dreams from REM sleep, the average adult will experience three to six REM periods every night, and the longest period lasts anywhere from 20 to 40 minutes. Yet the average adult recalls only about four to six dreams *a month*! Even those dreams that *are* remembered, unless written down, tend to be evanescent, their details quickly fading as we make our way through the day.

If the function of dreams depended on our remembering them once we awaken, then the overwhelming majority of dreams would be wasted. Or consider how some morning you awaken with no memory of having dreamt, only to have something you see or do later in the day bring back the memory of a dream from the night before. Now, some half-dozen hours after the dream took place, does the dream acquire a function because you suddenly remembered it? That doesn't make much sense. What does make sense is that whatever the function of dreaming is, it takes place "live," while the dream is actually unfolding.

As for any interpretation of those few dreams that we do remember, think of the last ten or twenty dreams you've remembered. How many of these did you have interpreted? Setting aside the issue of who did the interpreting, with what methods, and whether the interpretation was ultimately accurate, chances are your answer will be zero, or nearly so. In fact, children and teenagers (who sometimes remember significantly more dreams than the average adult) rarely have their dreams interpreted; and, for generations, people throughout the world have lived their entire lives without having a single dream interpreted. And if any other species are capable of dreaming, they certainly don't have their dreams interpreted. The evolutionary function of dreaming simply cannot depend on dreams being interpreted.

This is why it's important to distinguish between the uses we *choose* to make of the dreams we remember—for interpretation, personal growth, inspiration, or entertainment—and the biological or adaptive function of dreams. Over millennia, hundreds of ideas have been put forth to explain the nature and function of dreams, and dozens of ideas have been proposed since the discovery of REM sleep in the 1950s. All of these started from the assumption that dreaming has a biological function. But what if it doesn't? What if dreams are nothing more than a meaningless by-product of the sleeping brain?

Next to the dream theories of Freud and Jung, the activation-synthesis hypothesis proposed by Allan Hobson and Robert McCarley of Harvard Medical School is probably the most widely known theory of dreaming. In a pair of articles published in 1977, Hobson and McCarley presented a model of dreaming based on the neurobiology of REM sleep[3] while simultaneously carrying out a frontal assault on Freud's psychoanalytic dream theory.[4] The model was overtly anti-Freudian—the authors focused on how their hypothesis was at odds with Freud's dream theory, while largely glossing over areas where the two theories agreed.

In a nutshell, activation-synthesis proposes that dreaming is triggered by the "largely random" firing of giant neurons in the pontine reticular formation (PRF) of the brainstem. The PRF is a diffuse network of neurons located in the pons between the forebrain (the bumpy outer portion of the brain that people usually picture when they think of the brain) and the spine. The PRF plays a role in the regulation of REM sleep, and Hobson and McCarley proposed that the firing of these giant neurons during REM sleep stimulates the visual cortex while simultaneously initiating the rapid eye movements that REM sleep gets its name from.

According to the activation-synthesis hypothesis, the forebrain responds to this stimulation by attempting to construct a narrative that explains these visual sensations. The dream is, according to Hobson, the result of the forebrain "making the best of a bad job in producing even partially coherent dream imagery from the relatively noisy signals sent up to it from the brainstem."[5]

Although Hobson and McCarley's articles paid much attention to the brainstem "activation" half of the model, discussion of the "synthesis" half—the elaboration of the brainstem stimulus by forebrain structures—was relegated to a single paragraph. As time went on, Hobson seemed to take a certain pleasure in emphasizing the random aspect of dreaming and then basking in the often outraged responses this provoked. As a result, the activation-synthesis model is largely seen today as proclaiming dreams to be random and meaningless.

Somewhere between this Scylla of random, meaningless dreams and the Charybdis of dreams carrying messages from the gods or the unconscious, there slowly grew a sometimes inchoate movement proposing cognitive and emotional functions for dreaming. Even Hobson and McCarley, in their 1977 activation-synthesis paper, suggested that there could be "a functional role for dreaming sleep in promoting some aspect of the learning process."[6] In contrast, Francis Crick, co-discoverer of the structure of DNA, suggested in 1983 that REM sleep served the function of reverse learning, whereby the brain erases memories replayed in the dreams we never remember—"We dream in order to forget."[7] Thus, according to Crick, the worst thing we can do is try to remember our dreams!

The next thirty years saw a flood of dream function theories; we apologize to the many whose models we haven't the room to discuss. But most of these claim one or more of the following: (1) the function of dreams and REM sleep are one and the same; (2) dreams help us solve problems; (3) dreams have an evolutionary function; (4) dreams play a role in emotional regulation; (5) dreams have no adaptive or biological function; and (6) dreams have a memory function. Let's look at each of these ideas.

REM Sleep = Dreaming

Despite all evidence to the contrary, many people continue to use the terms *REM sleep* and *dreaming sleep* interchangeably, and some even refer to nonREM sleep as *dreamless sleep*. Thus people will say, "This research shows that REM sleep does XYZ" and then conclude that *dreaming* does XYZ. Similarly, a study showing that depriving rats of REM sleep leads to hypothermia goes on to conclude that dreaming may help keep the brain warm.

When a recent study *in mice* found two genes (called Chrm 1 and Chrm 3) that were necessary for REM sleep, the headlines that followed included "Your dreams come from two genes," "Genes regulate how much we dream," and our favorite, "Removing genes that code for dreams may allow scientists to stop nightmares." What these headlines should have said was "Mouse REM sleep depends on Chrm 1 and Chrm 3 genes," which in itself is an important discovery. Less exciting, perhaps, but infinitely more accurate.

Moreover, as we saw in previous chapters, REM sleep is a *physiologically* defined sleep stage, whereas dreaming refers to the *subjective experience* people have while sleeping. When we ask if dreams have a function, we want to know whether the *experience* of dreaming—of flying, meeting a dead relative, or discovering a new room in your childhood home—has a function beyond what may be going on in the underlying REM and nonREM sleep physiology (for example, increased release of the neuromodulator acetylcholine in the brain). Additionally, dreaming is not restricted to REM sleep, and so the function of dreams cannot be the same as that of REM sleep.

Dreams Help Us Solve Problems

One of the more straightforward ideas about why we dream is that dreaming helps us find solutions to our personal problems. Proponents of this idea often point to famous discoveries made in people's dreams, including Elias Howe's invention of the sewing machine, August Kekulé's discovery of the benzene ring, or Paul McCartney's writing of "Yesterday." We'll have more to say about the wonderful links between dreams and creativity in Chapter 11, but these celebrated examples also serve to highlight the rarity of such events given the *billions* of dreams being dreamt worldwide every night.

Great discoveries and breakthroughs aside, studies have shown that dreams only rarely contain practical solutions to real-life problems. This is not to say that dreams can't help us grapple with important problems or concerns. People will sometimes make a decision, plot a course of action, or reconsider a previous plan based on a dream. But such insights or solutions typically emerge only later, while thinking about the dream *after awakening*. Yes, dreams can lead to fantastic discoveries and insights, but actual problem solving within dreams clearly occurs too seldom for it to be the reason that dreaming evolved.

Dreams Have an Evolutionary Function

One interesting way of thinking about dream function is to consider what adaptive advantage dreaming may have conferred on our ancestors. In 2000, the Finnish philosopher and cognitive neuroscientist Antti Revonsuo published a provocative and widely debated evolutionary model of dreaming; he proposed that dreaming evolved as a mechanism for simulating threatening events and rehearsing possible means of avoiding or surviving them.[8] Since then, his threat simulation theory (TST) of dreams has been the focus of multiple studies and detailed critiques.

Taken as a whole, empirical support for the TST has been mixed. Although it is true that much dreaming can be conceptualized as containing psychologically or physically threatening material, realistic life-threatening events occur infrequently in dreams. Furthermore, only a small percentage of dreams depicting realistic threats to survival contain effective avoidance responses. For example, in a study of 212 adult recurrent dreams (selected because Revonsuo viewed repetitive dreams as exemplary cases of threat simulations), Tony and his colleagues found that a third of the recurrent dreams did not contain threatening events.[9] In fact, 80 percent of the threats identified were fictional in nature or very unlikely to occur in waking life (for example, having bathroom walls disappear, being visited by a ghost, or flying unaided over a body of water). The research team also found that successful avoidance responses occurred in fewer than 1 in 5 threatening recurrent dreams, and in fewer than 1 in 25 recurrent dreams overall. Moreover, 40 percent of recurrent dreams containing a threat ended with the threat being fulfilled, and another 40 percent ended with the person waking up before the threat was resolved one way or the other.

Similarly, many nightmares can be seen as simulating failures rather than any form of adaptive response. Thus, while many dreams contain various forms of perceived threats, few show evidence of effective and realistic behavioral responses to those threats. This is a key finding because Revonsuo's model argues that it is not so much the simulation of threats that was biologically adaptive, but rather the rehearsal of successful *responses* to these threats.

In 2016, Revonsuo and his colleagues put forth an alternative social simulation theory of dreaming, proposing that the function of dreaming is to simulate and thereby strengthen "the social skills, bonds, interactions, and networks that we engage in during our waking lives."[10] As with TST, the social simulation theory of dreaming postulates that dreamt simulations improved the survival and reproductive success of our ancestors. The social simulation theory has been critiqued on theoretical and empirical grounds, but it's too early to tell how well this alternative evolutionary model of dreaming will fare in comparison to TST. But in both cases Revonsuo and his team put forth well-articulated and testable models of dream function, a feat that is relatively rare in the field.

Dreams Help Regulate Our Emotions

There has been, in the past decade, a veritable explosion in the number of neurobiological studies supporting the idea that sleep—REM sleep in particular—plays a key role in emotional processing. In the overwhelming majority of these studies, however, there is no data on dream content. Because of this, inferences from these studies remain speculative as to what, if anything, dreaming contributes to sleep's role in modulating daytime emotional functioning.

That said, many clinical theories of dream function proposed over the past forty years have focused on the very idea that dreams play a role in emotional regulation. One model, reminiscent of Freud's view of dreams as the guardians of sleep, views dreaming as a kind of thermostat, tasked with containing emotional surges or regulating the dreamer's mood across successive periods of REM sleep. According to this theory, when dreams succeed in their overnight modulation of emotions, there is a pre- to post-sleep improvement in the dreamer's mood; or as Bob's mother used to say, "Things will look better in the morning."

Other models have proposed that by combining negatively toned dream imagery with muscle paralysis, REM sleep dreaming carries out an adaptive "desomatization" function that uncouples emotions from their underlying physiology. Indeed, blood pressure, heart rate, and breathing patterns are often uncoupled from ongoing dream emotions.

More recently, neurocognitive models of dysphoric dreaming—bad dreams and trauma-related nightmares—propose that one function of dreaming is to reduce or extinguish fear-based memories by allowing fear stimuli to be experienced in novel and emotionally varied circumstances.

Likely the best-known modern clinical theory of dreaming is that of the late Ernest Hartmann, a former psychiatrist and professor at Tufts University, whose contributions helped further our understanding of dreams and nightmares. Based on his work with trauma victims and accepted findings on REM sleep physiology, Hartmann proposed[11] that dreaming is a form of "nighttime therapy" that helps weave emotional concerns as well as traumatic events into existing memory systems, all within the "safety" of sleep. Dreams, according to Hartmann, accomplish this function by creating connections between new and old memories that are broader and looser than connections made during wakefulness. His model also considers emotions as the instigators of dream content (as opposed to emotions occurring in reaction to the dream) and that the dreamer's emotions guide the connections made during dreams.

Another core feature of Hartmann's theory is that dreams can be viewed as containing a "contextualizing" image, a kind of picture-metaphor that stands in for key emotional issues. Hartmann gives the example of how a dream image of being swept away by a tidal wave captures the feeling of being overwhelmed, a feeling attributable not to a past experience with tidal waves, but to an unrelated trauma such as being in a fire or the victim of sexual assault. According to Hartmann, as the dreamer's emotions change over time, so too does the associated contextualizing image, facilitating emotional adaptation. Although Hartmann's model is relatively vague in describing the mechanisms through which dreams would exert their therapeutic effects, it offers an intriguing and clinically informed conceptualization of dreams, particularly those related to trauma. Rosalind Cartwright, another pioneer of sleep and dream research, offered similar ideas on the emotion-regulating function of dreaming in her studies of recently divorced men and women.[12]

Taken as a whole, these models suggest that dreaming, or at least REM sleep dreaming, does play a role in regulating negative emotions. We'll explore this idea further in the next chapter, but for now, we'd like to take a step back and ask, "What if each of these theories has some element of truth to it?" Could it be that dreams sometimes help us solve problems, but at other times they provide a unique environment in which to rehearse social interactions, learn to avoid threatening situations, or process emotions? We think the answer to this question may well be yes.

Dreams Have No Function

Maybe, maybe not. A number of clinicians, philosophers, and dream researchers believe that dreaming serves no adaptive biological function at all. Owen Flanagan, a philosopher at Duke University, sees dreams as spandrels—unintentional side effects of evolution, like the sound of the heartbeat or the color of blood—that might nonetheless be useful when recalled.[13] David Foulkes, who has conducted some of the more intriguing studies of dreaming in children, maintains that dreaming does not have a biological function.[14] The eminent dream researcher and theorist William Domhoff, who has spent decades compiling well-founded research showing that psychologically meaningful information can be extracted from people's dream reports, has nonetheless argued that dreaming was not conserved by natural selection although dream recall, like the capacity to generate music, can play a significant role in people's lives.[15]

Similarly, some researchers working on the relations between sleep and memory have suggested that while dreams may *reflect* underlying processes of memory reactivation and evolution, the dreams themselves serve no function. For some of these researchers and many others, dreaming is nothing more than a meaningless epiphenomenon of the sleeping brain.

Dreams Have a Memory Function

It's clear to us that dream content isn't random, and that dreaming isn't merely an epiphenomenon. To be honest, however, we both went through periods during which we doubted if dreaming truly served a biologically adaptive function. Like the authors of many of the models presented earlier, we hadn't fully anticipated key discoveries about sleep's role in memory processing, nor about the role that REM and nonREM dreaming might play in these processes. We saw in Chapter 5 that both REM and nonREM sleep contribute to memory evolution. Sleep enhances some memories while allowing others to be forgotten. It processes both emotional and unemotional memories. It enhances some memories in great detail while extracting gist, discovering patterns, or inspiring insight from others. And it does this selectively, saving and evolving memories that the brain calculates will be of greatest future use. We firmly believe that any reasonable model of dream function must take all of these findings into account.

In the model we propose in the next chapter, dreaming is a form of sleep-dependent memory processing, albeit a phenomenologically complex one, that extracts new knowledge from existing information through the discovery and strengthening of previously unexplored associations. In doing so, dreams rarely replay active concerns directly or offer concrete solutions to them. Rather, they identify and strengthen associations that in some way embody these concerns and that the brain calculates may be of use in resolving them or similar concerns, either now or in the future.

What Function Does Dreaming Serve?—Take 2

When Erin Wamsley joined Bob's lab in 2007, she wanted to document the connection between dreaming and sleep-dependent memory processing. Erin had just completed her PhD thesis on REM and nonREM dreaming at the City University of New York, under the direction of John Antrobus, who had made his name studying daydreaming and mind wandering, but who also studied dreaming. While at the university, Erin worked on a study run by her classmate and future husband, Matt Tucker, who was working on sleep dependent memory consolidation. When she came to Bob's lab, she wanted to put the two together. By the time Erin joined the lab, researchers knew that sleep supported memory evolution, processing new memories from the day before. But they didn't know if or how dreaming contributed to this process.

Bob had reported in a 2000 *Science* paper that learning a new task could affect dream content. Using the classic computer game Tetris, he had shown that participants learning the game dreamed about it at sleep onset in no uncertain terms, reporting relatively accurate images such as "Tetris shapes floating around in my head like they would in the game, falling down, sort of putting them together in my mind."[16] But that wasn't all. With his colleague Margaret O'Connor, Bob was able to study Tetris dreams in five patients with amnesia. Their amnesias had been caused by accidental damage, mostly carbon monoxide poisoning, to a structure deep within the brain known as the hippocampus. The hippocampus is critical for learning and recalling recent events. Without a working hippocampus, patients couldn't remember what they had for breakfast or where they had gone that afternoon. David Roddenberry, one of Bob's students, sat with each of these patients while they played Tetris for a total of 7 hours over three days. Each night he sat by their beds, monitoring their sleep and awakening them to collect dream reports. Before bed each night, all of the patients reported having no memory of playing Tetris. In fact, they had no memory of ever meeting David. (One night a patient asked him, "Why are you in my bedroom?")

Yet despite having no memories of playing the game, three of the five patients saw Tetris images in their dreams; for example, one reported seeing "images that are turned on their side. I don't know what they are from—I wish I could remember—but they are like blocks."[17] Known as the Tetris Effect, this phenomenon has gained considerable notoriety, spawning everything from its own Wikipedia page to a signature lunchbox that Bob uses to a PlayStation virtual reality game bearing the same name.

Still, Bob had never combined dreaming and memory evolution into a single experiment. That was Erin's job.[18] She designed a virtual maze and had participants explore it, trying to learn its layout. Then she let them take a 90-minute nap. After their nap, she asked them whether they remembered dreaming about the task and then tested them in the maze again.

The results were astonishing. Participants who had no memory of dreaming about the task took, on average, 1½ minutes longer to find their way out of the maze after their naps, while those who reported that they had dreamed about it found their way out 2½ minutes *faster* than before their naps. With these findings in hand, Erin took the plunge. She repeated the experiment, this time waking participants during their naps to collect actual dream reports. Bob had always been afraid to try this, fearing that it would interrupt any ongoing memory processing. (He needn't have worried. A study carried out some eight years later would show that waking participants even five or six times in a night had no effect on memory evolution.[19])

When Erin identified those participants with dream reports related to the task, she found that they showed almost *ten times* more improvement after their naps—averaging 91 seconds faster at retest—as compared with the participants who reported no related dreams and were just under 10 seconds faster at retest. Clearly, those who were dreaming about the maze task were showing more sleep-dependent improvement. Plus, Erin had the dream reports. They were better than a fuzzy fMRI image showing increased activity in the hippocampus—she could look directly at what the sleeping brain was imagining as it presumably enhanced this newly stored information.

The answer from one participant was: "I was thinking about the maze and kinda having people at checkpoints, I guess [there were no people or checkpoints in the maze], and then that led me to think about when I went on this trip a few years ago and we went to see these bat caves, and they're kind of like, maze-like." Another subject reported: "[I was] looking for something" in a maze. And another recalled "just hearing the music" that played in the background while they had been exploring the maze.

This was not good. These dreams were not going to help participants enhance their memories of the maze's layout, which could have helped them navigate through the maze faster after they awoke. And yet these were the very participants who showed the greatest improvement. The sleeping brain was both enhancing its memory of the maze layout *and* creating related dreams. But the dreams, while predicting the subsequent improvement in task performance, could not have been directly contributing to that improvement. These dreams must be doing something else, serving some other function. But what?

We saw in Chapter 5 that the sleeping brain performs multiple forms of memory evolution. It selects recent salient memories for nocturnal processing, prioritizing emotional memories but also processing unemotional ones; it stabilizes and strengthens some memories while extracting rules and gist from others; and it integrates new memories into older, preexisting knowledge networks. Fortunately, the brain is great at multitasking and can probably carry out multiple forms of memory processing at the same time. For example, after Erin's participants fell asleep, the hippocampus presumably replayed and strengthened its memories of the paths they followed earlier in the maze. (It's known that the hippocampus does exactly this in rodents.) But this leaves the rest of the brain free to deal with other aspects of the evolution of these memories, such as how to file them.

Should we file these new memories under "ways to make a quick fifty bucks"? How about under "computer games I've played" or "the time I got separated from Mom in Costco and thought I was lost"? Or, as these dream reports would suggest, should we file them under "looking for lost items" or "exploring bat caves" or "I really hate techno rap music"? This is not a trivial question. Our brains store immense amounts of information in an unbelievably complex collection of interlocking neural networks where related memories are physically connected to one another so that activation of any memory in the network will tend to activate others in that network. How the brain decides to file new information—into exactly which preexisting networks to link a new memory—determines how and when this new information will be available during subsequent wakefulness. For instance, will the maze come to mind when you need to make some quick money or when you're playing a new video game or exploring a cave?

On the other hand, how your brain files these new memories also determines what will come to mind when you try to navigate the maze again, after your nap—whether you will recall the caves or getting separated from your mother. More important, it allows the brain to discover and strengthen creative links between these memories. Perhaps some strategy you learned while exploring a cave will help you in the maze task, or conversely, maybe something you learned from the maze task will help you next time you're down in a cave. Your brain suddenly realizes, hey, exploring mazes and caves is really the same thing. And, of course, that's exactly what Erin's participant dreamed: "that led me to think about when I went on this trip a few years ago and we went to see these bat caves, and they're kind of like, maze-like." It's a perfect example of the function of dreaming we suggested earlier—the extraction of new knowledge from existing information through the discovery of unexpected associations.

WHY DO WE HAVE TO EXPERIENCE OUR DREAMS?

Must we actually dream to get these admittedly profound benefits of sleep-dependent memory processing? In their maze learning paper,[20] Erin and Bob concluded that dreaming was a *reflection* of brain processes underlying sleep-dependent memory processing. But they didn't argue that dreaming per se—the actual conscious experiencing of the dream—had anything to do with this memory processing. Dreaming, they seemed to imply, was nothing more than a spandrel, an inconsequential side effect of important memory processing being carried out by the sleeping brain. This brings us back to the idea that dreams are epiphenomenal, with no real function. Or, as William Domhoff had argued almost twenty years earlier, maybe "Dreams were not conserved by natural selection to be problem resolvers, or anything else for that matter, but they nonetheless can be used to understand our unfinished emotional business because they happen to express our conceptions of our preoccupations."[21]

Today, both of us reject this idea. As we'll detail in the next chapter, we believe that dreaming, like waking consciousness, offers two advantages beyond what can be accomplished by nonconscious brain processing. Dreaming creates narratives that unfold in our minds across time and allows us to experience the thoughts, sensations, and emotions engendered by those narratives. Dreaming, like waking consciousness, allows us to imagine sequences of events, to plan, to plot, to explore. Even when a problem doesn't inherently require the development of a narrative—for example, figuring out whether adding two odd numbers always produces an even number—we nonetheless create narratives to help us solve them. We "think out loud" about it, "run it through our mind," and sometimes go through a series of "steps" as we solve it.

In his book *The Feeling of What Happens*,[22] Antonio Damasio, a professor of neuroscience, psychology, and philosophy at the University of Southern California, argues that this creation of narratives is one of the greatest powers of consciousness. We cannot, he argues, construct narratives outside of consciousness. And without the capacity to construct them, we wouldn't be able to recall the past, imagine the future, or plan ahead—abilities that make us human.

Being able to imagine and plan for the future is similarly critical for several forms of sleep-dependent memory evolution; and to do so, the sleeping brain needs to dream. By dreaming, the brain creates conscious narratives that imagine and explore a host of possibilities in a way that other, nonconscious forms of sleep-dependent memory processing cannot. We must dream if we want to perform these functions of sleep.

Damasio also argues that the subjective experiencing of emotions (what he calls feelings) is at the core of consciousness, critical for even common everyday decision-making. Even when making "totally rational" decisions, we rely on our emotional readout to confirm that we have made the right choice. Damasio offers the example of a woman with a genetic disorder that led to the destruction of her brain's amygdalae. Without her amygdalae, she could not experience fear or anger. As a result, she had never learned the signals of potentially unpleasant and frankly dangerous situations that the rest of us learn to read—and depend on—as children. On a simple gambling task, she couldn't learn which choices were bad bets, even though she saw them go bad time after time. Because she couldn't *feel* the wrongness of her choices, she couldn't *learn* their wrongness.

From this perspective, it's not surprising that emotions are so prevalent in dreams. If we accept Damasio's conclusion that the experiencing of emotions is critical for the evaluation of even apparently straightforward situations, then it's clear that we need to experience emotions in our dreams if we are to evaluate them—to understand what they mean for us.

Thus, emotionally engaged narrative dreaming is required for the full exploration, evaluation, and strengthening of novel associations relevant to our ongoing concerns. This process, in a nutshell, is the biological function of dreaming. We'll spell out this function and how it works in the next chapter, where we describe a new theory of dream function: NEXTUP.

Chapter 8

NEXTUP

IN THIS CHAPTER, WE PRESENT A NEW MODEL OF dream function that explains why a human brain must dream to carry out critical components of its sleep-dependent memory evolution function. The model is called NEXTUP—network exploration to understand possibilities. It was initially developed and named by Bob and then vastly improved with help from Tony. We detail NEXTUP's defining features, provide support for each of them, and discuss their implications. By the end of the chapter, you'll have an even better understanding of why brains dream. Let's begin.

NEXTUP AND THE EXPLORATION OF WEAK ASSOCIATIONS

NEXTUP proposes that dreaming is a unique form of sleep-dependent memory processing that extracts new knowledge from existing memories through the discovery and strengthening of previously unexplored weak associations. Typically, the brain starts with some new memory, encoded that day—maybe an important event, a discussion overheard at work, or something related to a personal concern—and searches for other, weakly associated memories. These can be from the same day, or they can be older memories from any time in the dreamer's past. The brain then combines the memories into a dream narrative that explores associations the brain would never normally consider. In doing so, NEXTUP searches for and strengthens the novel, creative, insightful, and useful associations discovered and displayed in our dreams.

Bob measured the brain's preference for weak associations during REM sleep in a study published back in 1999.[1] He used a cognitive test called semantic priming, developed by James Neely at Yale twenty years earlier. It's a clever test. Participants sit in front of a computer screen as a series of words and non-words, such as *right* or *wronk*, are flashed on it. Their task is to respond to each of them by pressing a key labeled "word" or one labeled "non-word." At the end, Bob calculated how fast and how accurate participants were when responding to the words and non-words. But that's not the whole story. Before each of these targets was displayed, another word was flashed on the screen for a quarter-second. Depending on the semantic relationship between this "prime" word and the target word (when it *was* a word), people responded more or less quickly.

You can see examples of how this test might play out in Figure 8.1. Participants identify the word *wrong* faster when it's preceded by a strongly associated word like *right* than when it's preceded by a weakly related word like *thief.* And they respond faster in both of these examples than when it's preceded by a completely unrelated word like *prune.* How much faster someone responds is a measure of their semantic priming. When Bob tested participants during the day, he got exactly the results he expected—strong primes like *right* produced three times as much priming as weak primes like *thief.*

But what does it mean? Every time you see a word, your brain activates circuits that allow it to remember the sound and meaning of the word. But it also activates memories of related words. Not only does this activity enable you to understand the word better, but it

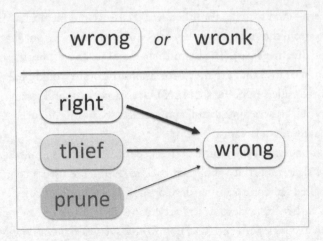

Figure 8.1. Semantic priming in wake.

also prepares the brain for what might be coming next. And the more strongly it activates the memory of a given related word, the faster and more reliably you will be able to identify that word when it does come next. That's exactly what we're measuring here. When you respond to *wrong* a lot faster after a strong prime like *right* (compared to an unrelated word like *prune*), it means your brain strongly activated the target word *wrong* in response to it. Bob's results suggested his participants' brains were activating strongly related words three times more effectively than they did weakly related words.

Bob was able to run this test very quickly; he could get participants through the entire test in just 2 to 3 minutes. This is much faster than the time required for the brain to become fully awake, and for its levels of neuromodulators like serotonin and noradrenaline to shift to waking levels. By testing participants as soon as they woke up, he was able to ensure that their brains' neuromodulator levels were still close to where they had been before waking up. He tested them right after waking them from REM sleep in the middle of the night, and the results, shown in Figure 8.2, were better than he could have hoped for. Priming produced by strongly related words dropped by 90 percent,

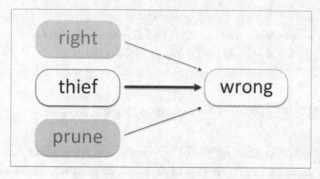

Figure 8.2. Semantic priming in REM.

while that produced by weak primes increased more than twofold. When participants were awakened from REM sleep—and presumably while they were in REM sleep just a few minutes earlier—their brains were activating weakly related words eight times more effectively than strongly related words.

When our brains dream, this preference for weak associations helps explain why so many of our dreams lack any transparent connection to the dominant thoughts, feelings, and events of our day. Even when connections are obvious, the usefulness of a dream usually isn't. But this is exactly what NEXTUP predicts—weakly associated networks are being explored to understand possibilities. The brain is searching more widely than during wakefulness, going through less obvious associations, and digging for hidden treasures in places it would never consider while awake. In the glare of day—when our brains are dealing primarily with new incoming sensations and the balance of neurotransmitters in our brain is optimized for processing the here and now—the usefulness, or "rightness," of these newly found associations might be incomprehensible. But that's fine. We don't need to understand why our brain chose these associations. We don't need to know whether the associations used to construct a given dream were useful. We don't even need to remember the dream. All the important work was done while we slept. Associations were discovered, explored, and evaluated while we dreamed, and if our brain calculated that some of them were indeed novel, creative, and potentially useful to us, then it strengthened them and filed them away for later use.

NEXTUP and Dream Bizarreness

Another consequence of NEXTUP's preference for weak associations is the prevalence of bizarreness in dreams. The presence of bizarreness is so striking that many earlier dream theorists felt obliged to explain it. At one extreme, Freud's disguise-censorship hypothesis explained bizarreness in dreams as part of an intentional process of disguising forbidden wishes that weren't safe to express directly in dreams. At the other extreme, Hobson described this bizarreness as a result of the forebrain almost randomly slapping images and concepts together, "making the best of a bad job."[2] But from our perspective, it's simply a predictable consequence of weak, and hence unexpected, associations being incorporated into the dream narrative.

Even when the relevance of the dream to "day residue" (the unfinished thoughts and feelings from the day) is overt and straightforward, bizarreness still creeps in. In Bob's first faculty position, at the University of Massachusetts Medical Center, he had the unpleasant task of helping to teach the so-called dog lab. This long-since-abandoned lab was the medical students' introduction to death, albeit in the guise of a laboratory study of cardiovascular function. When students arrived, they were confronted with anesthetized dogs and a lab manual describing how to insert catheters into veins, measure intravenous blood pressure, inject drugs, and more. Near the end of the lab, they would cut through the skin and muscles of the dog's chest, use a buzz saw to go through the rib cage, and apply drugs directly onto the pumping heart. Bob was too squeamish and couldn't bear watching the students cut through the ribs. He left this task for his colleagues to supervise. The first night that he taught the lab, he had a dream:

I was in the dog lab again, and we had just cut open the dog's chest. As I looked down, I suddenly realized that it wasn't a dog; it was my five-year-old daughter, Jessie. I stood there dumbfounded, not understanding how we could have made such a mistake. And as I watched, the edges of the incision drew back together and healed without a hint of a scar.

Waking from the dream, Bob told his wife about it, and she suggested that the dog lab had clearly aroused his fears of mortality. And where were these fears greatest? For his child Jessie, of course. But Bob disagreed. That's not what it felt like to him. To him, the dream seemed to ask the question, "If it's okay to do this to a dog, why isn't it okay to do it to Jessie?" Of course, both explanations were reasonable.

This is a classic example of what dreams like to do. At the time, Bob's brain took an emotional event from his day and replayed it with an entirely improbable and bizarre rescripting. Obviously, this dream wasn't designed to improve his ability to perform the surgery. Instead, while his brain dreamed, it searched through his memory networks for weak and potentially useful associations: The dog and Jessie were both small and helpless; he felt responsible for them both; he didn't want either of them to die; he loved them both; or all of the above. Finding multiple links to Jessie, his brain built her into the dream. But why? Not to answer a question, and not to solve a problem. Rather, to do just what NEXTUP evolved to do. The brain asked "What if?" and watched its own emotional and cognitive response, observing how this response affected the dream narrative. The intensity of that response and the way it influenced the rest of the dream told the brain what it needed to know: This association, Jessie and the dog lab, was a valuable one. Something was uncovered about the fragility or sacredness of life that was important, something worth marking and strengthening and keeping available for the future. Once these connections were strengthened, the brain's job was done. Whether Bob remembered the dream when he woke up or not didn't really matter.

Of course, not all dreams are as straightforward as this one. It's often a real stretch to identify any connections between our dream content (or at least what we can remember of it when we wake up) and what happened that day or at any point in the more distant past. And sometimes, what the brain finds is of no use whatsoever.

It's also important to keep in mind that, as we'll see in Chapters 10 and 12, dream imagery—much like spoken language, films, and stories—can be figurative and metaphorical in nature. Dream content sometimes dramatizes current concerns and other meaningful life events without displaying any concrete elements from them.[3] In either case, trying to find a connection after we wake up—between, on the one hand, the novel and unusual associations generated by NEXTUP within the rich, imaginative context of a dream and, on the other hand, the myriad thoughts, feelings, and events experienced during wakefulness—is not only tricky, but prone to all sorts of errors. We'll have more to say about this later.

NEXTUP: Understanding Possibilities

We've discussed the concept of network exploration in some detail, but we haven't talked about the idea of understanding possibilities (representing the last two letters of NEXTUP). In Chapter 5 we discussed sleep-dependent memory evolution and the brain processes that allowed for the stabilization, enhancement, and integration of new memories, as well as the extraction of gist and discovery of rules within these memories. In all these instances, there was an implicit understanding that the brain was serving the function of improving our memories by producing more useful memory representations. This function is akin to what's known as *convergent thinking* (see Figure 8.3, left panel)—seeking to identify the single correct answer to a question, reaching its conclusion logically and, in the end, leaving no ambiguity.

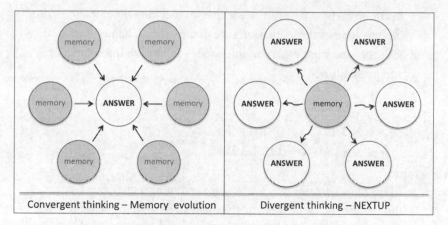

Figure 8.3. Convergent and divergent thinking.

In contrast, NEXTUP is more akin to *divergent thinking* (Figure 8.3, right panel), which works in a creative, free-flowing manner to generate a number of insights and potential answers to an initial question. NEXTUP proposes that our dreams allow us to explore network connections in order to *understand possibilities*. You can think of it as similar to the true goal of education—not to cram facts into our brains, but rather to open us to the unexplored possibilities embodied within those facts, showing us the many ways they can be used and not just one specific way.

Think back to Bob's dog lab dream. No specific problem needed to be solved, nothing for which a "correct" answer could be found. There was only a recent memory of a deeply disturbing event that Bob somehow needed to understand. We have no direct access to what his brain cells were doing while he was dreaming, only to his report of the experience they created. This means we can't somehow measure what NEXTUP was doing at the neural or network level during the crucial segments of his dream. But Bob's dream report suggests that his brain was seeking an understanding of the event by exploring possibilities. And while NEXTUP did not need Bob to remember his dream when he woke up, it's clear that his thinking about it after waking up—one of those spandrels we discussed in Chapter 7 that add value beyond what evolution intended—did lead to interesting ideas and possibilities.

NEXTUP AND THE FELT MEANING OF DREAMS

Many of our dreams may feel strange and meaningless, but a surprising number of them seem to engender in us a strong sense of their importance. Why does this happen? If the function of dreaming doesn't require our remembering them, and if we remember so few of them in any case, why should they feel so meaningful when we do? (Indeed, this idea of meaningful dreams is seen across all cultures and across thousands of years.)

We know that the brain is specifically searching for weak associations. This means it is exploring associations that, under normal circumstances, it would reject as somewhere between uninteresting and just plain ridiculous. When we're dreaming, the brain must shift its bias toward scoring associations as potentially valuable when it normally wouldn't. It needs to give itself a little push if it's going to decide that any of the weak associations incorporated into its dream narratives are meaningful and useful.

It's a little like the sixties, when people were dropping acid and having profound "acid insights" along the lines of, "When you flush the toilet, *everything* goes down!" They would tell you this, wide-eyed in awe at their amazing insight, then get a bit sheepish and say, "It meant more than that; it really explained *everything*."

In fact, the feeling that dreams have meaning is not just a little like the sixties. It's likely identical. Bear with us here for a few sentences on neurochemistry. Pharmacologically, lysergic acid diethylamide (LSD) works by activating serotonin receptors, including serotonin

1A receptors, which in turn can block the release of serotonin in parts of the brain. All the weirdness of LSD—the hallucinations and acid insights and everything else—may be a direct consequence of this bio-chemical blockade of serotonin release. This obviously isn't the nor-mal state of affairs in the brain. But there is one time every day when serotonin release is completely blocked, and that's during REM sleep.

We dream in both REM and nonREM sleep, but the most bizarre, emotional, and unlikely dreams—and arguably those that seem most meaningful to us—occur in REM sleep. The reduction in serotonin levels during nonREM sleep (relative to wakefulness) and the com-plete cessation of its release during REM sleep may serve the impor-tant role of shifting the brain's bias toward assigning more value than it otherwise would to those weak associations activated during dream construction. This chemical action may be the grease that enables these potentially useful new associations to slide into our repertoire of valu-able insights, and in so doing produces a felt sense of meaningfulness.

NEXTUP and the Stages of Sleep

Back in Chapter 4, we introduced the multiplicity of sleep stages that occur each night, both REM and nonREM, and briefly discussed how they differed physiologically. Then, in Chapter 5, we discussed memory evolution during sleep and noted how some forms of mem-ory processing seemed to occur primarily during REM sleep. As you will see later in this chapter, the form of dreams also differs across sleep stages—by now a well-established fact. But what has not been addressed before is whether the *function* of dreams differs across sleep stages. Is the function of dreaming in REM sleep different from its function at sleep onset or during other stages of nonREM sleep? We believe so. Let's start by reviewing what we know.

First of all, there is something fundamentally different about how the brain dreams in REM and nonREM sleep. As we pointed out in

Chapter 4, our bodies are functionally paralyzed during REM sleep. It turns out this is necessary to prevent us from acting out our dreams. In REM sleep behavior disorder, this paralysis breaks down and individuals literally act out their dreams, hitting their bed partner, jumping out of bed, or shouting and gesticulating wildly. But in nonREM sleep there is no paralysis, and yet we don't act out our dreams. In REM sleep, when we're dreaming, the brain activates the motor cortex, the region of the brain that controls our movements, just as it would if the events in the dream were really happening. As a result, the brain needs to block our muscles from responding to signals from the motor cortex lest we act them out in reality. Because we have neither the paralysis nor the acting out of dream movements during nonREM sleep, the brain must not be activating the motor cortex as it does in REM sleep. Why we have evolved two distinct patterns of brain activation during REM and nonREM dreaming is a mystery. But these patterns tell us that when brains dream in REM and nonREM sleep, they're doing things differently—and they suggest that the two types of dreams are performing different functions.

There are also differences in the chemical neuromodulators being released in the brain. These chemicals control how nerve cells communicate with one another; at the whole brain level, they essentially act to switch the software running the brain. You just learned that serotonin can affect a dreaming person's sense of how significant a weak association is. When serotonin release is blocked during REM sleep, it leads to an increased sense of wonder and importance for whatever weak association happens to be found. Serotonin release isn't fully blocked during nonREM sleep, so this bias toward favoring weak associations will be diminished. But that's just fine, because the brain isn't looking for weak associations during nonREM sleep.

Bob's semantic priming experiment showed that our normal preference for strong associations is replaced by one for weak associations during REM sleep. This effect is probably due to a second neuromodulator, noradrenaline, that's also being shut off during REM. Noradrenaline is the brain's version of adrenaline; one of its many functions is to focus our attention on what's right in front of us. You've probably noticed that when you're under pressure and your adrenaline levels skyrocket, you don't want to think about a bunch of unlikely alternatives to what you are trying to do. You're so focused, almost nothing can change your course. The disappearance of noradrenaline from the brain during REM sleep makes it easy for your brain to wander among its weak associations. Bob saw evidence for this difference between REM and nonREM sleep in his semantic priming study. When he awakened participants from nonREM sleep, the weak primes, like *thief,* which had been so powerful during REM sleep, surprisingly had no effect at all (Figure 8.4). They didn't seem to be activated any more than the totally unrelated primes. Now, strong primes once again produced the biggest effect. It appears that only strong associations are being activated during nonREM sleep.

What does this tell us about the function of dreaming in different sleep stages? Before we can answer this question, we must digress from our discussion of sleep and dreams to a discussion of wakefulness, specifically to dreaming's cousins, daydreaming and mind wandering.

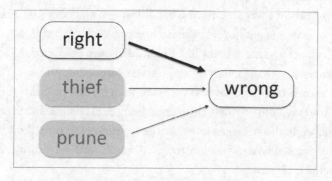

Figure 8.4. Semantic priming in NREM.

NEXTUP AND THE DEFAULT MODE NETWORK

In the last quarter of the twentieth century, two brain imaging techniques were developed that would dramatically change our understanding of how the brain works. Positron emission tomography (PET) and functional magnetic resonance imaging (fMRI) have allowed scientists to look at brain activity while people are performing a whole range of mental tasks and create detailed 3-D maps of that brain activity. Using these new techniques, we have learned which brain areas are "turned on" during all sorts of mental activity—from staring at geometric patterns to viewing emotional pictures, reading, and memorizing word lists to having out-of-body experiences.

To produce these maps of brain activity, individuals are rolled into the center of a huge doughnut-shaped machine. Pictures of their brain activity are then taken, first while they're just lying there resting and again while they're doing some task. If we subtract the activity patterns seen while individuals are resting from the patterns seen when they're performing the task, we get a picture of which brain regions are activated by the task—a map, if you will, of which parts of the brain actually perform the task. The use of fMRI to map brain activity in real time was an amazing breakthrough, and scientists quickly began mapping out dozens of brain functions.

As more and more brain imaging studies were published, it slowly became clear that something strange was happening. While turning on its own specific set of brain regions, each mental task also turned off other regions. At first this seemed quite reasonable. But as time went on, it also became clear that the regions being turned off were the same, no matter what task the participants were performing. And this made no sense.

Over time, however, Marcus Raichle, along with his colleagues at

Washington University in St. Louis, realized what they were seeing.[4] Scientists had been assuming that the activity pattern seen during quiet rest reflected the activity of a brain not doing anything. In retrospect, this was obviously a foolish assumption. Our brains are always thinking about something. Because of this, the brain areas that turn off whenever we start to carry out a mental task are the regions that do whatever the brain does when we're "not doing anything." Together these regions make up the default mode network (DMN), whose discovery has helped us appreciate just how true it is that the brain never rests.

When we look at the brain regions that make up the DMN, we find a sub-network that monitors the environment for important changes, watching out for any danger. Keeping us safe is probably one function of the DMN. But we also find a sub-network that helps us recall past events and imagine future ones, another that helps us navigate through space, and yet another that helps us interpret the words and actions of others. And these are the mental functions associated with mind wandering. Much of mind wandering involves hashing over the events of the day or anticipating and planning future events. Indeed, such planning has been proposed as a function of mind wandering.[5] So it's perhaps not surprising that mind wandering is associated with increased activity in the DMN.[6] This appears to be a second function of the DMN.

The DMN is not a static structure, however. It changes based on what you've been doing earlier. Bob and his colleague Dara Manoach looked at how activity in the DMN changed after doing one of Bob's favorite tasks: his finger-tapping task, which we saw in Chapter 5 involves learning to type the sequence 4-1-3-2-4 as quickly and accurately as possible.[7] Young participants get a lot better in just a couple of minutes of practice, but then they plateau. A period of rest in the same day doesn't make them any faster, but if they get a night of sleep and then try again, they become 15 to 20 percent faster. It's another example of sleep-dependent memory evolution.

When Bob and Dara had participants learn the task while having their brains scanned, with periods of quiet rest before and after the training, they found that brain regions involved in performing the task were talking to each other more during the quiet rest after training than during the quiet rest before training. The DMN, which is normally measured during such periods of quiet rest, was altered by performing the task. And more important, the more the DMN was altered, the more improvement participants showed the next day. It was as if this new DMN activity was telling the brain what to work on once it fell asleep.

Indeed, much of the DMN is also activated during REM sleep, suggesting that the term *daydreaming* may be more appropriate than we thought. William Domhoff and his colleague Kieran Fox have gone so far as to suggest that dreaming, or at least REM sleep dreaming, constitutes a brain state of "enhanced mind wandering."[8] More recently, Domhoff has proposed that the neural substrate of dreaming lies within the DMN.[9] When you put it all together, you get an exciting extension of our NEXTUP model. Whenever the waking brain doesn't have to focus on some specific task, it activates the default mode network, identifies ongoing, incomplete mental processes—those needing further attention—and tries to imagine ways to complete them. Sometimes it completes the process shortly after the problem arises, making decisions without our ever realizing it. But at other times it sets the problem aside after tagging it for later sleep-dependent processing, either within or without dreaming. Several dream theories have suggested something like this—that dreaming helps us address areas of concern in our lives. The DMN might provide the mechanism for identifying these concerns, thereby determining what's NEXTUP.

NEXTUP and Dream Function in Different Sleep Stages

With this added insight from the DMN, we can return to our question of how the function of NEXTUP might vary in different sleep stages. These differences are likely the greatest when looking at sleep onset. The hypnagogic period is a unique link between pre-sleep mind wandering and early sleep dreaming. A "fracture point" often takes place in sleep-onset mentation, wherein rational waking thoughts— inevitably about waking concerns or incomplete mental processes— shift into hypnagogic dreams.

Perhaps, then, it's not surprising that Silvana Horovitz at the National Institutes of Health outside Washington, D.C., has found that the DMN is active throughout the hypnagogic period.[10] She also saw a dramatic increase in brain activity in visual processing regions after sleep onset. Other features of these dreams give additional support to the idea that these dreams have a unique role in NEXTUP. Hypnagogic dream reports from the sleep-onset (N1) stage of sleep are dramatically shorter than other nonREM and REM dream reports. They're often clearly related to the thoughts you were having immediately before falling asleep, frequently evolving smoothly, if unpredictably, from those thoughts. Hypnagogic dreams are usually less bizarre and much less emotional than dreams from later in the night, and they often lack two features that are almost always present in other dreams; namely, self-representation and narrative structure. Much of the time these dreams are just unusual thoughts, or a random geometrical pattern, or a simple picture, like a landscape or a face.

This finding does not feel like fertile ground for the work of NEXTUP. Instead, these brief hypnagogic dreams appear to extend the DMN's work into the sleep period, identifying and tagging current concerns for further sleep-dependent processing, and perhaps then beginning to identify associated memories for later consideration. But the very briefness of such dreams suggests that they can do little more than tag these memories, leaving more extensive processing for later in the night.

NEXTUP saves its best work for REM sleep. Compared to non-REM dreams, REM dreams are longer and more vivid, emotional, and bizarre, and they have more complex narratives. In addition, when people try to identify the waking sources of the content within these dreams, they report distinctly fewer episodic memory sources—memories of actual events in our lives that we can fully bring back to mind, essentially allowing us to relive the original event. For example, if you saw flying saucers in a nonREM dream, you might identify its source as a related episodic memory, saying, "Oh, those flying saucers looked just like the pizza I had for dinner last night." In contrast, in a REM dream, you'd be more likely to say, "Oh, they looked just like pizzas; I *love* pizza," thereby identifying a general *semantic* memory (I love pizza) instead of a specific *episodic* memory (I ate pizza last night). This example aligns with how we imagine NEXTUP working in REM sleep as it tries to use the simulated world of the dream to generalize from these memory sources and create a more integrated understanding of their meaning and importance (Figure 8.5).

By comparison, N2 dreams are shorter and less emotional, bizarre, and vivid. But perhaps the most telling difference is that the waking sources of N2 dream content tend to be more recent and more episodic—what you had for dinner, what your partner told you at dinner, who washed the dishes, and so on—and do not arise from less-

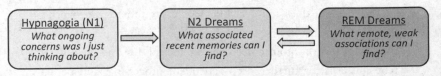

Figure 8.5. The function of NEXTUP by sleep stage.

specific "semantic" memories such as what you like to eat, what you often talk about with your partner, and which household chores are yours. The memory sources used in constructing N2 dreams thus lie between the immediate pre-sleep sources of hypnagogic dreams and the very loose, associative links to semantic memories seen in REM dreams. The function of these N2 dreams is probably intermediate as well. Although REM sleep appears to be seeking weak, often unexpected, remote associations that might be usefully related to memories of unresolved concerns from the day, N2 dreams appear to search for more obviously related episodic memories from the recent past.

Presumably, this logic would hold for all sleep-dependent memory processing, both within and outside of dreaming. Anna Schapiro, a former postdoc of Bob's now at the University of Pennsylvania, made this argument in a 2017 paper. She characterized the role of nonREM sleep (without reference to dreaming) as "an opportunity to recap the details of the day's events, providing additional exposure to information that was recently acquired from the world,"[11] and that of REM sleep as facilitating the "exploration of cortical networks containing long-term memories,"[12] a description that matches the definition of NEXTUP as network exploration.

This separation of REM and nonREM functions provides a rationale for the normal sequence of sleep stages across the night. Each night begins with N1, moves to N2 and N3, and then moves to REM before cycling between N2/N3 and REM for the rest of the night. As the night progresses, nonREM decreases and REM increases, allowing the brain to seek out weaker and weaker associations and our dreams to become more and more bizarre.

This evolution of dreams across the night can be seen even on a shorter time scale in the hypnagogic period. In another study from Bob's lab, using the arcade game *Alpine Racer II*, Erin Wamsley found that some dreams had strong, direct relations to the game, featuring unambiguous depictions of the game specifically, or of skiing in general; other dreams had weaker, more indirect relations to it, containing sensations, locations, or themes related to the game. When reports were collected at the start of the hypnagogic period, within 15 seconds of sleep onset, dreams were eight times more likely to show direct incorporation of the game than indirect incorporation. But after just 2 minutes of sleep, the rates of occurrence had become the same for both direct and indirect incorporations. Another group of participants were allowed to sleep for 2 hours before any reports were collected. After the 2 hours, they were awoken and then allowed to fall back asleep. They were then awoken again within 2 minutes and their sleep-onset reports collected. Now the ratio of indirect to direct incorporations was five times higher than for those collected right at the start of the night.[13]

Interestingly, Stuart Fogel at the University of Ottawa in Canada had participants practice the Nintendo game *Grand Slam Tennis* before collecting eight sleep-onset dream reports that night. Overnight improvement seemed to depend on how similar dream incorporations were to the actual game for the first four of these sleep-onset dreams, but not for the last four.[14] Perhaps only the earlier, more direct incorporations successfully tagged the game memories for NEXTUP.

This all makes sense. Whether we look at how pre-sleep activities are linked to dreaming in the hypnagogic period or how they're connected to dreaming in all the sleep stages across the rest of the night, dreaming appears to play an important role in how memories are selected and how they subsequently evolve across the night.

NEXTUP AND INSOMNIA

If you've ever suffered from even brief bouts of insomnia, you probably know the sensation of your brain seeming to run at full speed, rehashing everything worrisome and incomplete from your day, when you just want to relax and fall asleep. Why does your brain do this? Indeed, anxiety—be it stress, concerns, or apprehensions—is a major cause of insomnia. (On a given night, it might instead be due to excitement, such as when "visions of sugarplums danced in their heads.")

Why do all of these thoughts and images come crashing into our mind just as we're trying to fall asleep? Based on what we've just said, the answer is simple; the brain is using the sleep-onset period to tag current concerns—incomplete processes—for later processing during sleep.

Although the increasing rates of insomnia around the world may well reflect increased stress and worrying, we think there is another contributor—smartphones and earbuds. Take a look at people walking down the street, driving in their cars, eating alone in restaurants and cafes. Not so long ago, these people wouldn't be doing anything else. Their minds would wander and they would daydream; their DMN would be active, and, although they were totally unaware of it, they would be tagging recent memories for processing later that night. But as first the Walkman and then the iPhone came to dominate our free time, the DMN has slowly been squeezed out of our daily lives. Maybe all those worries come crashing in at bedtime because it's the only time we've left the brain to perform the critically important task of identifying and tagging memories for later processing. Maybe you can't have your iPhone and sleep, too.

Is NEXTUP for the Birds?

Let's assume for the moment that both dogs and babies experience some kind of dreaming. If, as NEXTUP proposes, dreaming serves the evolutionary function of exploring and coming to understand previously unconsidered possibilities, what does it do for them? Presumably it wouldn't have been maintained throughout mammalian evolution if it only came to serve a function in adult humans. But much as the content of dogs' and babies' dreams must be immensely less complex than in our own dreams, so the functional benefits of their dreams must also be diminished. If, as Dan Margoliash suggested, songbirds dream of singing and, as Matt Wilson suggested, his rats dream of running mazes, how does this jibe with the concept that dreaming supports our network exploration to understand possibilities?

Even with their diminished cognitive abilities and experience, rats can still benefit from NEXTUP. Even though we can't know whether or what rats dream, let's assume for the moment that they do, and that the activity researchers see in their brains during sleep *does* reflect their dreaming about mazes. In a variation on Matt Wilson's rat maze experiments, Anoopum Gupta, along with his colleagues at Carnegie Mellon University in Pittsburgh and others in Minneapolis, created a maze with two "T" choices (Figure 8.6).[15] During training, Gupta repeatedly placed a rat at the starting point in the maze. The rats learned that they had to go from there to the first T-choice (dashed arrow, T1), turn right, and proceed to the second T-choice (dashed arrow, T2). Some rats were then taught that they had to turn right at T2 (black arrows) to get food (F) on the right-hand side of the maze. If they turned left (white arrows), they found the left-hand food bowl empty. Other rats learned the opposite, that they had to turn left and not right.

Then, halfway through the test session, all of the rats had to switch directions to get food. If they had been trained to turn left, now they had to turn right, and vice versa. A third group of rats had to learn to

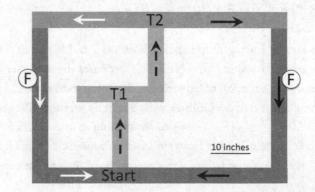

Figure 8.6. Design of the maze for training rats.

turn both ways, constantly switching which way they turned, turning left one time and right the next. All the while, Gupta was recording from brain cells in their hippocampus that reflected where they were in the maze, so he could follow their path either by looking at them or simply watching which cells were firing at any given time.

When the rats subsequently slept, he continued to record from these brain cells and saw them firing in the same order as when the rats were actually in the maze. Sometimes they fired in the same order as when they turned left, at other times in the order for turning right. But sometimes they fired in a totally unexpected order, as if they were going all the way across the top of the maze, from the upper left, past T2, and on to the upper right. The "place cells" in their brain were acting as if the rat was taking a path it had never taken in real life. If we continue to assume for the moment that the rat was dreaming, then it was truly carrying out network exploration to understand possibilities. Granted, it doesn't seem like a human dream; but within the limited cognitive capacities of rats, the narrative was probably equally provocative and unexpected. The rat was considering a possibility that it had never explored in waking. We think the same would be true for the songbirds. And newborns.

MANY OF THE DETAILS OF NEXTUP have been proposed before—some of them hundreds of years ago—and we've discussed several of these contributions in earlier chapters. But NEXTUP brings together a variety of innovative neuroscientific ideas and findings, offering concise explanations for a broad range of sleep- and dream-related findings. Its neurocognitive and neurobiological foundation allows us to extend NEXTUP to other mammals (as well as newborns), creating a broader, multifaceted, and developmental conceptualization of dreams and their potential functions. NEXTUP also provides an overarching context for the following chapters, starting with our discussion of dream content in the next chapter. A summary of the model can be found in the Appendix presented at the end of the book.

Chapter 9

THE MISCHIEVOUS CONTENT OF DREAMS

THE NEXT TWO CHAPTERS ARE ABOUT THE CONTENT of dreams and, as you may well expect, there is much to discuss. In Chapter 4 we talked about the myriad approaches that researchers take to collect dream reports. Whether we're studying the formal properties of dreams, the idiosyncratic content of individual dreams, or comparing different groups of dreamers, it all begins with the collection of dream reports. Only after this step can the real fun begin. With the right set of dream reports in hand, the list of questions we can ask is almost endless:

+ How long are the reports?
+ How many people, places, and things are in them?
+ How visual, auditory, and tactile are they?
+ How bizarre and emotional are they?

We can ask how these answers vary for different groups in the general population:

+ How do dreams differ between men and women?
+ How do they differ between adults, teenagers, and children?

We can ask about specific subpopulations:

+ How are dreams different in different cultures or from different centuries?
+ How are they different in people with sleep disorders like insomnia or sleep apnea?
+ How are they different in people with psychiatric disorders like schizophrenia or depression, or neurologic disorders like amnesia or Parkinson's disease?
+ How do they change during pregnancy?

And we can ask how dreams vary from night to night:

+ How are they affected by the time we go to sleep, or the amount of sleep we get, or who's in bed with us?
+ How do they differ across the night?
+ How are they affected by our thoughts and actions the day before? By stressors and hassles in our life? By what we ate?

Then, if we have sleep laboratory recordings, we can ask an entirely different set of questions:

+ How do the features of dreams differ in REM and non-REM sleep?
+ How are dreams different at the start of a REM period compared to 15 minutes into it?
+ Can we tell whether someone is dreaming from their EEG patterns? Can we tell how vivid or emotional their dream is?
+ What's going on in the brain during lucid dreaming, flying dreams, or nightmares?

These are the types of questions that we and our fellow dream researchers are always asking. They're our bread and butter. And they allow us, over time, to build up a picture of what dreams look like and of how what's happening in our brain affects them.

To answer these questions, we must collect dream reports and then extract the specific information we need—and that, in turn, requires a scoring system. When it comes to analyzing dream reports, the devil is truly in the details, and there is no shortage of scoring systems for getting to the answers.

In their 1979 book *Dimensions of Dreams*, Carolyn Winget and Milton Kramer catalogued nearly 150 dream content scales: they included everything from measures of ego integration to dream vividness to castration wishes.[1] The number of dream rating systems has probably doubled since their book was published, both in response to new research questions and to ever-evolving dream theories. But over the past fifty years, one coding system has stood head and shoulders above the rest and has probably been used in more studies than the next ten scoring systems combined.

HALL AND VAN DE CASTLE'S DREAM SCORING SYSTEM

The Hall and Van de Castle dream scoring system, developed in the 1960s by Calvin Hall and Robert Van de Castle, is the most well known, best validated, and perhaps most extensive dream scoring system ever designed. Although Hall was trained as an animal behaviorist, his research interests during his tenure at Western Reserve University gradually turned to dreams. Hall became particularly fascinated by how patterns of dream content observed in a long series of dreams could be used to infer aspects of the dreamer's personality, core conflicts, and personal concerns. In 1953 (the same year REM sleep was discovered), Hall published *The Meaning of Dreams*,[2] a popular book in which he presented an innovative cognitive theory of dreams that described dream images as "the embodiments of thoughts." But these are not just any kind of thoughts. Dreams, according to Hall, reflect the conceptions we have of ourselves, other people, the world, and our inner conflicts. His pioneering work helped advance the "continuity hypothesis" of dreaming, a widely held view that dreams reflect the dreamer's current thoughts and concerns as well as recent salient experiences.

Van de Castle, on the other hand, began with an interest in extra-sensory perception (ESP) and spent much of the late 1940s and early 1950s in the Parapsychology Laboratory at Duke University, where Joseph and Louisa Rhine carried out what are generally considered the most famous scientific studies of ESP ever reported. But Van de Castle's interest shifted toward dreams. Teaming up with Hall, he carried out studies of how home dreams differed from laboratory dreams and how dream content varied across REM periods. In a little-known side project, he also investigated dream telepathy, with Hall as the "sender" and Van de Castle as the "receiver"—but we'll say more about that later.

In a hallmark study, Hall and Van de Castle analyzed 5 dream reports collected from each of 100 male and 100 female college students—a thousand dream reports in all—in exquisite detail. Based on this work, they published *The Content Analysis of Dreams* (1966), a book that would revolutionize the scientific study of dream content.[3] In this 320-page manual, available online,[4] the authors described rules for scoring a wide range of dream features, including characters, settings, objects, and actions. They scored friendly, aggressive, and sexual interactions; successes and failures; good fortune and misfortune; and several emotions. The Hall and Van de Castle (HVC) coding system, as it came to be known, also included rules for parsing the information within each of these content categories in fine detail. Aggressive social interactions, for example, could be scored as a function of who initiated the aggression, whether it was reciprocated, as well as its severity—ranging from covert feelings of hostility to verbal threats and all the way up to murder. Similarly, the category of "characters," which included people, animals, and mythic figures, was further divided based on the number of characters and the gender, age, and identity of each one. The identity of a character could be, for example, an immediate family member, in which case it would be further subcategorized into mother, father, sister, brother, and so on. In the end, the HVC system allowed for an immensely detailed and mostly objective coding of dream reports, laying the foundation for fifty years of scientific studies of dream content.

Of course, for a given study, we probably wouldn't want to score the dream reports on all of these categories and their finer subdivisions. We might only be interested in the total number of characters in a dream, or whether the people in a dream are known or unknown to the dreamer. We might just want to know about the frequency of friendly social interactions, or the proportion of dreams that contain negative emotions. Again, it all depends on the question we're asking. The HVC scoring system even allows researchers to ask—and answer—important questions about dreaming without actually collecting any dream reports at all. This is done by using dream reports collected and scored by other dream researchers for completely different reasons. DreamBank[5] is an online database of over 20,000 dreams, many already scored according to the HVC system. A search engine allows scientists to search through these dream reports for specific words or phrases, as well as for specific HVC variables. A search can be limited to exploring Hall and Van de Castle's original data set of 1,000 dreams, can look at the long dream series of individuals from different backgrounds, or might focus on specific subgroups of dreamers, such as men, women, teenagers, Vietnam veterans, or the blind.

Using the HVC coding system, Hall and several other researchers showed that it was possible to extract psychologically meaningful information from the dreams of individuals unknown to them. The system has been similarly used to analyze the content of dreams of noted individuals, including Kafka, Jung, and Freud. The HVC system has even been used to study fictional dreams. Together with a team headed by Christian Vandendorpe, a professor of French literature at the University of Ottawa, Tony helped create a French website[6] featuring almost 1,500 literary dreams from over 400 authors. This database presents the context in which each dream occurs, and it also details the dream's significance within the work, its interpretation (if any is provided by the author), and of particular interest to us here, many of the HVC content variables described earlier.

After scoring the dream reports from the specific group researchers are interested in, the group's HVC scores are compared to normative values, such as those in *The Content Analysis of Dreams*, or to dream-content profiles from other specific populations, cultures, individuals, or time periods. The observed content patterns (for example, the settings, characters, or social interactions), together with these comparisons to other populations, allow researchers to identify key features of the dreamer's inner and outer waking life. Dozens of such studies using this approach have successfully extracted impressively large amounts of high-quality information about people's personalities, concerns, and activities from their dream reports.

Hall and Van de Castle's rating scales are still widely used today. But that doesn't mean all is well in the study of dream content. As you'll soon see, the scoring of some of the more subjective dream elements—most notably bizarreness and emotions—is far trickier than many people imagined. And depending on how these dream features are defined and assessed, they can give rise to contrasting results.

NEXTUP AND FORMAL DREAM FEATURES

What do dreams look like? It's impossible for us to summarize the thousands of studies that have examined dreams from almost every conceivable angle, but we'll try to answer the most important questions people have about their contents and, in doing so, give you a reasonably accurate picture of what dreams are like. We'll start by examining the general characteristics of dreams, including the formal properties of dreams mentioned in Chapter 7. We'll save what actually transpires in our dreams for the next chapter, where we'll explore topics such as the themes people most frequently dream about and the content of recurrent dreams, sex dreams, and nightmares. We'll also go beyond the findings and percentages to show you the complexities and challenges involved in figuring out what dream content is all about.

Based on the description of NEXTUP in Chapter 8, you should be able to make some predictions about dreams—both their formal properties and specific content—even if you've never recalled a single dream in your entire life. NEXTUP predicts that your dreams will usually contain sensory perceptions, a narrative structure, and emotions, and they'll have you as a player in the dream. They will juxtapose weakly associated concepts and events that you would not normally think of as going together, often giving your dreams a bizarre flavor. And their content will be related to your current concerns, usually with some connection to recent daytime thoughts, feelings, and events, but it will also include connections to older memories worth exploring in the context of these ongoing concerns. Let's check out these predictions.

FORMAL DREAM PROPERTIES IN DIFFERENT SLEEP STAGES

As you'll soon see, all the formal properties of dreams suggested by NEXTUP are indeed found in dreams, as well as others not necessarily predicted by our model. Some, if not all, of these features can be found in most dreams, especially in REM dreams. But as we discussed in the last chapter, many dreams, especially nonREM and hypnagogic dreams, contain few of these features, or none at all. Trying to sort all of these features for every sleep stage gets quite complicated. In general, our descriptions will be either for laboratory dreams from N2 and REM—where most of our recalled dreams come from—or for at-home dreams, where the sleep stages simply aren't known. With this caveat, let's look at NEXTUP's predictions one at a time.

Sensory Imagery

This is, without a doubt, the most striking feature of dreams. Whether they contain visual, auditory, tactile, olfactory, vestibular, or other sensory modalities, dreams are typically intensely vivid and compellingly realistic. In fact, except for those rare occasions when we dream lucidly, we always believe our dreams to be real while we're in them, and we only come to realize their illusory nature after we wake up. In general, all of us can picture things in our mind while we're awake—the face of a friend, the sound of a whistle, the heat from a fire—but for most of us, such waking images are a pale imitation of either real life or our dreams. Still, not all dreams contain such imagery. When all sleep mentation reports, however brief, are included in the analysis, about 10 percent of REM dreams and as much as 30 percent of nonREM reports turn out to lack any sensory imagery.[7] Instead, they're just thoughts. We don't know much about these dreams because they've generally been ignored by dream researchers, who historically haven't considered them to be "real" dreams. At sleep onset, such dreams probably serve to identify concerns for later processing. Whether the same is true for later thought-like dreams is unclear.

All sensory modalities are found in dreams, but they're obviously not all equally prominent. Most dreams with sensory perceptions contain visual images; sounds are reported only about half the time; and reports of smell, taste, and pain are each seen in less than 1 percent of reports. However, it's unclear whether these low numbers accurately reflect, for example, how often smells and tastes are present in dreams. In one study, Bob collected 900 reports from participants while they were awake, often during meals. And while there were 250 references to being at breakfast, lunch, or dinner, there were only 24 mentions of taste and 13 of smell—less than 15 percent of the number of meals reported. Clearly, most people don't normally report the tastes and smells they're experiencing. The same probably holds true for dreams. Still, we do know that only about a third of men and 40 percent of women recall, when asked, ever having experienced smells or tastes in their dreams.

Of course, not everyone's dreams are so highly visual. When either of this book's authors describes the visual nature of dreams during public talks, somebody invariably asks, "What about the blind? Do they have visual images in their dreams?" This is a great question, and one to which we have a definitive answer.

For a start, people who are blind from birth have no visual images in their dreams (although it's unclear how they could recognize them if they did). The same is true for those who become blind before the age of four or five. As you might expect, the dreams of the congenitally and early blind contain many more descriptions of how things feel, taste, and smell, including sensory details like the texture of their clothing, or the slight incline of the street where they were walking. They even report sensations like the warmth of the sun on their skin. But those who become blind after the ages of five to seven do, at least initially, continue to dream visually, although the frequency and clarity of these dreams can diminish over the years. People who become deaf show a similar phenomenon, with many reporting unusually vivid visual dreams. In both groups, the increased intensity of dream imagery in those senses that they still retain parallels similar gains during waking perception and imagining.[8]

Although the overwhelming majority of dreams appear to contain visual imagery, there remains the subtler but often-asked question of whether we dream in color or in black and white. The answer to this question is a bit complicated. In a recent online study,[9] respondents said they dreamed in color 50 percent of the time, in black and white 10 percent of the time, and couldn't remember the other 40 percent of the time. Among those who were born after the introduction of color TV, only one in two hundred said they always dreamed in black and white—that's eight times fewer yes responses than reported among those born before the advent of color TV. In contrast, 40 percent of participants in a 1942 study said that they always dreamed in black and white (including 51 percent of the men, but only 31 percent of the women). Apparently, before the introduction of color into movies and TV, people reported dreaming in black and white a lot more than they do now. But now, even these people almost all report dreaming in color at least twice as often as in black and white. Were they dreaming in color all along?

It's a great question. Possibly, the introduction of color into movies and TV made people more attuned to "technicolor" aspects of their vision, including in the movies playing nightly inside their head. But colors in dreams were reported at least as far back as the time of Aristotle, and even about half of the longer dream reports presented in Freud's 1899 edition of *The Interpretation of Dreams* contained explicit references to color. Maybe it was the popularity of black-and-white media, including black-and-white photography, movies, and television that flourished well into the 1940s and 1950s that originally led people to *think* that the images and scenes they recalled from their dreams were in black and white. After all, it was only in the early to mid-twentieth century that most people—including scientists—became interested in the question at all.

It makes sense to think that everyone who sees colors in real life can also see colors in dreams. But researchers know that we don't remember most of our dreams, and memories of the dreams we *do* recall start to fade the moment we wake up, so it's reasonable to think that these tenuous memories are more likely to be made up of core features of the dream—its setting, the people present, specific sequences of events—than secondary details like the colors of objects or the ambient temperature of the air. Thus, whether we recall colors from our dreams may depend more on what we're focused on while we're dreaming—and ultimately on what our brain is busy encoding while we dream—than on whether color is actually present in our dream imagery.

The answer, however, may not be black and white after all. It's even possible that while a dream as a whole appears to us in color, some objects or aspects within it do not—or the other way around. The beauty of this line of questioning is that, with some effort, you can literally *see* which of these possibilities is most often true of your own dreaming brain.

Narrative Development and Plot Continuity

NEXTUP proposes that dreaming enhances memory processing by allowing the brain to create narratives related to ongoing concerns and allowing the dreamer to react to them. Narrative sequences are so common in dreams[10] that we don't even think about it; it's just what dreams are. Dreams *could have* evolved to simply display visual images, like looking at pictures, but instead they have evolved to flow through time as if we were awake. In fact, Ed Pace-Schott, whom Bob worked with for over a decade, has argued that this aspect of dreams is hardwired into dreaming through the default mode network—you'll remember the DMN is involved with recalling past events and imaging future ones while we're awake—as a veritable "story-telling" instinct.[11] Bert States, who was a professor of dramatic arts at the University of California, has gone even further by suggesting that all forms of literature and theater come from what we observe in our dreams.[12] It's fascinating to consider the possibility that storytelling, literature, and theater are all waking derivatives of dreaming, serving the very function proposed by NEXTUP for dreaming. Certainly storytelling, literature, and theater do lead us to explore our own memories; and, at their most powerful, they help us understand new possibilities suggested in our own life.

A related but not so obvious dream property has to do with plot continuity: whether the plot of a dream continues coherently from start to end. Although dream plots do seem to be coherent from moment to moment, they rarely maintain that continuity over an entire dream. Some thirty years ago, Martin Seligman and Amy Yellen at the University of Pennsylvania described this *principle of adjacency*[13] as being similar to conversations at parties; each comment is relevant to the one before, but the topic drifts so quickly that participants often ask, "Wait, how did we start talking about this?"

Bob demonstrated the truth of this observation in a 1994 study of spliced dreams.[14] He took 22 dreams and divided them into 2 sets of 11. The first set, he left untouched. But he took the second set and cut each dream apart at the end of the sentence closest to the middle of the report; he then "spliced" each half of a dream to a half from another dream.

After shuffling the 11 resulting spliced dreams with the intact set, Bob asked five judges to guess which of the 22 dreams were spliced and which were intact. All told, the judges identified the dream reports correctly 90 out of 110 times, a rate that they would get by simply flipping a coin 110 times less than once in 100 billion tries. Clearly, most dream reports are coherent from one sentence to the next.

But what about from one end to the other? Bob took a second collection of eighteen dream reports, all at least twenty lines long, and divided them into two sets. This time he took just the first and last five lines from each dream. Again, for the first set of dreams, he simply put the two cuttings back together, the same as he had with the set of intact reports in the first experiment, except now the middle of each dream was missing. For the remaining set of dream reports, he created new spliced dreams by combining the beginnings and ends of different reports.

This time around, the results were totally different. Although the judges had been right over 80 percent of the time in the first experiment, in this one they were right only 58 percent of the time, not much better than the 50 percent you'd expect if they had flipped a coin. Still, a third of the reports—including three spliced and three intact reports—were correctly identified by at least six of the seven judges. How did the judges know? Each of the three spliced reports turned out to have different main characters in the two segments, which was enough to convince the judges that the spliced reports came from different dream reports. In contrast, each of the three intact reports had a person, place, or object that was present in both parts of the abridged report, providing an element of continuity that gave them away. So it looks like only about a third of dreams manage to maintain an obvious dream element from one end to another. In fact, for the six cases that almost all judges scored correctly, it was not some plot continuity (or lack thereof) that gave it away; it was the continuity of persons, places, and things.

What does this tell us about NEXTUP? Unlike a piece of classical music, there is no reprise at the end of a dream; and unlike a novel, the denouement does not fold back onto its beginning. This pattern makes good sense. Our dreams rarely come to neat endings. The most common end for a dream report is, "And then I woke up." The dreaming brain doesn't plot out whole stories. Indeed, the reduced levels of noradrenaline we discussed in the last chapter might prevent NEXTUP from staying on a single plot narrative for very long. Instead, NEXTUP stitches a series of memories and network explorations together, albeit keeping a principle of adjacency in operation. It's much like that cocktail party conversation, wandering from one topic to the next in an ever-evolving narrative, but always on the lookout for potentially useful new associations.

Self-Representation and Embodied Presence

When we dream, it's not like we're looking passively at pictures or watching a movie. It's more like participating in one of those multiplayer online role-playing games that Bob's sons play all the time. We're real characters in the ongoing dream events, a fact that's so common, it's often overlooked. But it is critical for NEXTUP and the dream process. Two Italian dream researchers, Pier Carla Cicogna and Marino Bosinelli, have identified eight distinct categories of self-representation in dreams.[15] Categories 1 through 5 range from not being present at all, as when watching a movie (category 1), to being fully present but only observing the events of the dream (category 4), to being an active participant interacting with other dream characters and objects (category 5). Categories 6 to 8 are stranger forms of self-representation. They include dreams in which the dreamer takes on the role of some other person, or even an object—in one case a photocopier—in the dream (category 6), or the roles of two people at once (category 7), or acts as both a participant and a lucid observer of the dream (category 8). Categories 6 and 7, while quite rare, are particularly interesting because they are also seen in some neurologic conditions when patients are awake.

All of these categories are seen in dream reports, but categories 4 and 5—present only as an observer and present as a participant—are the most common, and they are what dream researchers generally mean by "self-representation." As you might imagine, almost all REM dreams fall into one of these two categories—90 to 100 percent, depending on the study. In contrast, self-representation is seen in only two-thirds of nonREM dream reports and in somewhere between a quarter and two-thirds of sleep-onset hypnagogic dreams, depending on exactly how long after sleep onset the reports are collected.

In addition to self-representation, "embodied presence" also occurs in dreams. Cognitive scientists who support this intriguing concept argue that if we want to study how we come to understand the world around us and make decisions in it, it's not enough to study the brain and body as if they were separate from the environment. Rather, we must include the environment in our cognitive models of self. Thus, although classical cognitive science insists that we only have to worry about the neural representations of the outside world that our senses project to our brain, the principle of embodied presence argues that this scope is inadequate, and the physical environment must be seen as part of our cognitive machinery.[16]

Dreaming is an example par excellence of embodied presence. Although the brain produces dreams, the body clearly affects them, and bodily sensations are often incorporated into our dreams. More important, however, the external environment is replaced by a self-generated internal world. In a very real sense, our consciousness during dreams *enfolds* the virtual world that surrounds us; the world becomes a part of our consciousness.

By activating the neural networks that underlie both our sense of self and our conception of the world, the dreaming brain creates both us (the dreamer) *and* the dream world in which we find ourselves, launching us into an immersive and ever-evolving journey that we experience from a personal, first-person perspective. But in an often overlooked yet extraordinary process, the dreaming brain not only tracks how we respond to various situations in our dreams, it also tracks how the dream world *itself* reacts to our ongoing thoughts, feelings, and actions.

Together, the self-representation and embodied presence seen in dreams play a key role in NEXTUP, providing a near-perfect environment in which the brain can execute its dream functions. Although NEXTUP arguably could create usable narrative simulations from its exploration of memory networks without these features, such simulations would be like passively watching movies, and they would lack the verisimilitude created by this powerful combination. With these key features, the simulated world is consciously perceived and reacted to during a continually changing, dynamic interplay between the dream self and the dream world. This cycle of shifting dream world and subsequent dreamer response is what drives the construction of the dream narrative (Figure 9.1). And during this remarkable inter-

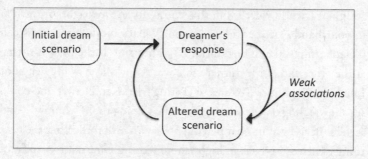

Figure 9.1. The narrative development loop.

play between our dream self and the rest of our simulated dream world, NEXTUP works its magic.

Bizarreness

Whether it's Bob's daughter Jessie dreaming of a duck in her bed or his own dream of a dog in his lab, dreams can show impressive bizarreness. After all, there's a good reason that people, about to share a freshly recalled dream, often begin with "I had the weirdest dream last night." But how unusual or strange *are* dreams?

Think back to your own dream experiences for a moment. How many would you say are truly bizarre? If Tony were to read fifty of your most recent dreams, do you think he'd come up with the same number as you would? What if your neighbor was scoring them, or your mom? Chances are, everybody's numbers would be somewhat different because of how they define *bizarre* and how much they know about you. The same is true in dream research.

For most people, including scientists, dream bizarreness encompasses things that are impossible (walking through walls, talking with the deceased, seeing a cat morph into a wolf) or improbable (running into a flock of sheep, being hit by a tsunami, getting back together with your ex). But dream bizarreness can be much subtler—maybe your best friend's voice is off, the season doesn't *feel* right, or the pen in your hand is shaped like a spoon. What's more, bizarreness can include things like uncertainties ("I'm not sure if the person sitting at the other end of the dinner table was Aunt Marie or my neighbor Julie"), incongruities ("We were visiting friends in Denver and could see ships sailing on the Pacific from their living room window"), and scene shifts ("I was in a bar playing pool with my brother, but the next thing I knew I was back in high school writing some math exam").

Bizarreness in dreams can thus refer to a wide range of events and experiences. This is one reason that over a dozen dream bizarreness scales have been developed, each with its own way of defining and scoring unusual elements in dreams. Moreover, while some of these scales focus on unusual *details* within dreams, others consider the dream experience *as a whole*. Unfortunately, dream researchers don't agree on which of these instruments should be used or what they tell us about the nature of dreaming. As a result, you'll find articles claiming that all dreams are highly bizarre and other articles concluding that most dreams are mundane. The truth, as you may well expect, lies somewhere in between. Here's what we know.

Approximately 75 percent of REM dream reports contain at least one occurrence of some form of bizarreness (a scene shift, incongruity, or uncertainty), but a considerably smaller percentage (less than 10 to 20 percent) contain three or more forms of bizarreness or even one distinctly impossible event.[17] We also know that dream bizarreness is somewhat less prevalent in reports from nonREM sleep (only about 60 percent) and that it shows up in only a third of sleep-onset reports. So, while most dreams show one form of bizarreness or another, some (including about a quarter of REM dream reports) contain no overtly strange or unusual features. In sum, unlike the claims made by opposing camps, not all dreams are bizarre, nor are most of them mundane. But the story doesn't end there. Four additional, often overlooked, aspects of dream bizarreness merit attention.

First, have you ever noticed that your dreams seem to be the longest and most vivid and intense when you wake up late on a weekend morning? Well, they are, and for two good reasons: because you're waking up later in the night (actually in the morning) and because you're probably awakening from unusually intense REM sleep. In a clever study, Erin Wamsley, who was then a graduate student, woke participants four times during the night to collect dream reports.[18] Participants were awoken twice from REM sleep, once early in the night and once late in the night, and twice from nonREM (N2) sleep, also once early and once late. When she analyzed the reports, Erin found that four dream features—length, dreaminess, bizarreness, and emotionality—all showed the same pattern. Dreams were longer, dreamier, more bizarre, and more emotional in REM sleep than in nonREM, as were both REM and nonREM dreams when they occurred later rather than earlier in the night. As a result, late-night REM dreams were the longest, dreamiest, most bizarre, and most emotional; the early-night nonREM dreams were the least bizarre. This is one reason that dreams recalled at home tend to be more bizarre than those collected throughout the night in the sleep laboratory.

Another aspect of dream bizarreness is a bias in the kinds of dreams people recall, as well as the kinds they choose to share with others. Research suggests that the presence of bizarreness in a dream facilitates its encoding, thereby making these dreams more likely to be recalled than their dull and unremarkable counterparts. But people's belief that most dreams are outlandishly bizarre also exists because these are precisely the dreams that we are most likely to share with one another. If, after waking up, you remember a dream in which you were sitting in your car, stuck in traffic, chances are you're not going to go out of your way to tell your friends about it. But if, while you were stuck in traffic, a gigantic eagle clasped the hood of your car in its talons and lifted you high above the city streets—giving you a breathtaking view of the city below as it ferried you to your destination with powerful flaps of its massive wings—and winked a large golden eye at you before flying off, *then* you might very well insist on telling everyone about your dream. The same goes for dreams shared in therapy. People almost never bring short, boring dreams to their analyst; they bring the long, perplexing ones that captured their attention and generated an interest in further exploration. And of course, these dreams are then the ones that make their way into the more clinically oriented dream literature.

A third reason we consider dreams bizarre is that we compare them to regular waking life. In fact, this is about the only aspect that virtually all dream bizarreness scales have in common. But is this the right comparison? Take the bizarre scene shifts that characterize so many dreams. When compared to normal waking reality, these shifts in location, perspective, or action are certainly bizarre. But what if we compared dreams to movies, or better still, to mind wandering, when our thoughts naturally shift and sway from one moment to the next and from one location to another. If, as suggested in Chapter 8, dreaming is governed by the default mode network (the same system that appears to underlie mind wandering), then such scene shifts are *exactly* what we'd expect to see in people's dreams.

Finally, dream bizarreness isn't a chaotic free-for-all; even the most unusual dream transformations tend to show certain constraints, forms of internal "dream logic." One study that Bob carried out with Allan Hobson and Cindy Rittenhouse[19] involved looking at bizarre transformations of objects and characters in dreams. Bob and his colleagues showed that transformed items usually remain within their class—objects remain objects, and characters remain characters. Thus a car may turn into a bike, and one person may morph into another (or even a mixture of several people); but objects don't turn into people, and people don't turn into lamps or tools or plants.

In the category of object transformations, even stricter constraints appear to be active. You can see them for yourself in Table 9.1. The table contains eleven objects mentioned in dream reports, numbered from 1 to 11 on the left, and the eleven objects, A to K, that they spontaneously changed into on the right. But items A to K have had their order scrambled. Item 1, the bag, did not change into object A, the bike. Try to match items 1 to 11 with their transformed objects, A to K. Answers are in the footnote on this page. In Bob's study, six judges tried their hand at matching them and made only 4 mistakes in the 66 attempted matches. Because there's only one chance in eleven of matching any one of the objects correctly by chance, the odds of getting only four wrong are vanishingly small. Thus, this facet of dream bizarreness reveals the presence of associative constraints operating within our dreaming brain, as NEXTUP would predict.

Notice that when one object suddenly changes into another, the

Table 9.1

BIZARRE OBJECT TRANSFORMATIONS

Try to match the dream objects in the left-hand column with what they turned into, in the right-hand column. Answers appear at the bottom of the page.*

1	bag	A	bike
2	bed	B	school bus
3	Boston home	C	intense combat video game
4	building	D	Georgia home
5	car	E	figures
6	cash machine	F	half-size bed
7	pool	G	building
8	car	H	burlap sack
9	city bus	I	beach
10	flowers	J	lion (really a bed)
11	statue of a lamb	K	car wheels and frame

* Answers: 1-H, 2-F, 3-D, 4-G, 5-A or K, 6-C, 7-I, 8-A or K, 9-B, 10-E, 11-J

new object is always similar to the old one. This is important. When NEXTUP explores associative networks, it could select a new object because it's *similar* to the original object or because it's *related* to the original one. But in every case in Bob's study, NEXTUP chose a new object based on its physical similarity to the old one. None of the objects turned into something that we've learned is related to it. The city bus, for example, didn't turn into a bus driver. If you're aware of the bizarre transformations in your own dreams, you'll intuitively know that changes like this never occur. It would appear that such transformations—which the dreaming brain could easily produce— aren't useful for understanding possibilities in our lives, and so they are rejected by NEXTUP as it constructs our dreams.

But it's also possible that the associations underlying some of these transformations are not as weak as they look to outside judges. Some may, in fact, seem perfectly reasonable to the dreamer. Consider one of Tony's dreams:

> *I'm back in the house where I grew up. As I enter my bedroom, I see my cherished teddy bear lying on the bed. As I move to pick it up, I hear someone enter my room. It's my uncle Romeo, whom I haven't seen in years. I give him a hug and then, as I pull away, am puzzled to find myself standing in the middle of a vineyard.*

This sudden shift in location fits with the ones we talked about earlier in Bob's spliced dreams study. Obviously, such a sudden shift could never happen while Tony was awake. But are the two locations as unrelated as Bob's study would suggest? Maybe not, once you learn that Tony's beloved teddy bear was a gift from his uncle Romeo, who came from Italy for a visit when Tony was a child, and that Romeo used to tease Tony's mother about North American wines, telling her that even the cheapest *vino* made from the grapes growing near his home in Italy was of better quality. And even if Tony hadn't remembered some of these details, his dreaming brain certainly did.

There are several conclusions we can draw from this discussion of dream bizarreness. First, most dreams are not what the movies and mass media make them out to be: they're not Felliniesque creations in which your mother zips by overhead on a trapeze as you skateboard uphill on water beneath her, while your brother—who is really a cat—watches from atop a glowing birdhouse. In fact, to experienced dream researchers, these kinds of accounts (often presented at the start of popular media articles on dreams) sound exactly like what they are: over-the-top stories made up to match people's wildest conception of dreams. Research on thousands of laboratory and at-home dream reports tells us that most dreams, especially REM dreams, are captivatingly strange, but not as outlandish as the media might lead you to think. In reality, dreams usually present us with convincing settings and characters, and it's often a conversation, situation, or unfolding of the plot we find ourselves in that is most strikingly odd. The same can also be said for many plays, novels, and movies.

Furthermore, when brains dream, they don't create a chaotic jumble of scenes. This is true even of people's strangest dreams, and even when the dream's associative processes and memory sources aren't readily apparent to the dreamer. (You'll find some amusing examples of this phenomenon in the next several chapters.)

But most important, all these findings make sense from the perspective of NEXTUP. Strange and unusual juxtapositions of content coming from different memory sources are just what we're looking for. They give your brain a chance to explore unexpected and weak associations within a convincing dream world, designed by your brain to elicit a range of reactions from you within the dream.

Emotions

The last prediction of NEXTUP we will discuss pertains to emotions. As you may remember, NEXTUP proposes that our dreams need to have emotions (or a "feeling of what happens") to allow the brain to interpret the potential value of the weakly associated memories that have been coactivated. Indeed, most dreams do contain emotions. We examined over a dozen studies of dream emotions and found that people rate 70 to 100 percent of their dreams as emotional. Curiously, when external judges rate dream reports, they typically identify emotions in just 30 to 45 percent of them, even when scoring exactly the same dream reports.[20] It appears that many of us tend not to explicitly mention our emotions in our dream reports, just like we usually fail to report smells and tastes.

But what counts as an emotion in the first place? Some models list five to seven categories of basic emotions; others list as many as twenty. If scientists can't agree on what emotions are found during wakefulness, they are unlikely to fare much better when it comes to dreams. In their seminal book on dream content, Hall and Van de Castle proposed just five categories of dream emotion: happiness, sadness, anger, apprehension, and confusion. They took this approach in part to get a high degree of inter-rater reliability when scoring dreams. In contrast, others have used fifteen categories of dream emotion: interest, exhilaration, enjoyment, surprise, distress, anger, disgust, contempt, fear, shame, shyness, guilt, excitement, jealousy, and anxiety. And this much longer list doesn't even include apprehension or confusion, two of Hall and Van de Castle's five categories. To make comparisons even more difficult, some studies don't measure discrete categories of emotions in dreams at all; instead, they just score the dream's overall emotional tone (for example, "generally positive or negative"). It can all get quite messy.

Thus, as with dream bizarreness, findings on dream emotions depend on what scales are used, who's applying them, and what types of dream reports (for example, those recorded at home on awakening spontaneously in the morning versus others recorded in the laboratory after forced awakenings from nonREM sleep). Averaging the scores across a dozen studies, we found that people rated their dreams as positive about three-fourths of the time when they slept in the laboratory, but only half the time when they slept at home. This last figure is almost identical to what some of these studies found when participants also recorded events from the day; people rated the emotions in their waking life as positive 51 percent of the time.

When independent judges scored whether the emotions in dream reports were positive or negative, they were uniformly about 25 percent more negative than when participants rated their own dreams. The judges rated laboratory reports as positive only about half the time (compared to three-fourths of the time when participants rated their own dreams) and scored at-home reports positive only a fourth of the time (compared to half the time when participants rated them).

A similar picture emerges regarding the intensity of dream emotions. On a scale of 1 to 5, participants on average rated the intensity of both positive and negative emotions at 3.2, just a bit higher than the middle value of 3. They gave the intensity of their waking emotions the same rating—an average of 3.3.[21] Not surprisingly, where participants are sleeping and who's doing the rating seem to matter here, too. In one study,[22] participants and independent judges rated the intensity of negative emotions similarly, but participants rated their own positive emotions as being more than twice as intense as did the judges. And in another study, participants rated the intensity of their positive emotions the same for home and lab dreams, but rated their negative emotions as almost three times more intense when they were dreaming at home.[23]

Despite all the differences in these findings, we can still draw some general conclusions. First, people do have emotions in their dreams, particularly in dreams with a basic narrative structure. But in general, everyday dreams aren't very intense; they average somewhere between mild and moderate, and they're no more intense than our emotions during the more important events from the day. Second, our dream emotions are well balanced between positive and negative in overall tone, and again they're not that different from what we experience when awake. Third, we don't do a very good job of reporting our emotions in our written dream reports. As a result, our dream emotions, especially positive ones, aren't that obvious to those who are reading our dream reports. Finally, the emotions in our dreams just don't seem to be strong enough to explain why we find dreams so important, and why we often feel compelled to share them with others.

TO SUMMARIZE WHAT WE'VE described in this chapter, dreams contain convincing sensory experiences embedded in a narrative story with moment-to-moment, but not beginning-to-end, continuity. They are characterized by bizarreness, nearly omnipresent feelings or emotions, self-representation, and an embodied sense of self. All of these features match what we would predict based on NEXTUP. But it's important to note that we've only been talking about the formal properties of dreams and not their actual content. In the next chapter, we turn our attention to the specific content of our dreams and ask what it implies for NEXTUP.

Chapter 10

WHAT DO
WE DREAM ABOUT?
AND WHY?

WHEN PEOPLE TALK ABOUT THEIR DREAMS, THEY rarely focus on their formal properties. Instead, they describe the story—the dream's setting, the people and objects present, and the overall storyline. In other words, they focus on the dream's specific content. In this chapter, we'll examine what people dream about in their everyday dreams as well as in the kinds of dreams that we're often more interested in, like recurrent dreams and nightmares. Then, we'll turn to the implications of these findings for NEXTUP.

EVERYDAY DREAMS

Our garden variety, everyday dreams are as different from one another as are the events in our waking life, but certain patterns and preferences are seen in what we dream. As we discussed in the last chapter, virtually all of our dreams with a basic narrative structure contain the dreamer as an active participant and are typically experienced from an embodied first-person perspective. Yet we're rarely alone in our dreams. Most dreams contain at least two other characters, and the people in our dreams—including ourselves—are usually involved in some activity, such as looking or walking, or some social interaction like talking with other dream characters. About half of our dream characters are familiar to us—relatives, friends, colleagues, or acquaintances—while the other half are unknown, including strangers and people identified solely by their occupational role—a policeman, doctor, or teacher.[1]

When we look at the gender of our dream characters, a peculiar finding emerges: women's dreams contain an equal proportion of male and female characters, but men's dreams contain twice as many male characters as female. Why this gender difference exists is still under debate, but it has been documented in numerous studies, across cultures, and even in the dreams of young girls and boys.

We also know that animals—from owls to tigers to Fido the dog— appear in up to 40 percent of children's dreams, but in only about 5 percent of adult dreams. But this finding seems to be more clearly a cultural distinction. In studies of several preindustrial and hunter-gatherer societies, adult dreams feature animals about five times more often than the 5 percent found in more urban populations,[2] presumably because people in those societies are closer to and interact more often with animals and their natural environments.

Another key feature of dreams is the interactions that take place between characters, including with the dreamer. According to Hall and Van de Castle, aggressive social interactions occur in a slightly greater proportion of people's dreams (46 percent) than do friendly interactions (40 percent). Furthermore, physical aggression is significantly more common in men's dreams than in women's—and women, in turn, are more likely than men to be victims of aggression in their dreams, mirroring differences seen in most cultures.

Looking at other features of dreams, we find that misfortunes— mishaps the character cannot avoid—are seen in about a third of all dreams. That rate is, sadly, seven times higher than the rate of good fortunes. But we succeed in handling difficulties in dreams as often as we fail.

Of course, our dreams take place *somewhere*. These settings are entirely familiar in only a third of our dreams, but that's still twice as often as the settings are entirely unfamiliar. In the other half of our dreams, the settings are vaguely familiar. Women's dreams take place indoors just over half the time, while those of men are more likely to occur outside, and here again, we don't know why.

At a more global level, the dreamer—or another character—is usually faced with some kind of problem. These can range from relatively minor difficulties—planning a course of action, trying to make sense of a situation, or finding a lost object—to serious physical or psychological dangers such as being lost, falling ill, facing interpersonal conflicts, dealing with environmental hazards, or fleeing from physical perils.

In these everyday dreams, the characters, interactions, problems, and setting are highly idiosyncratic. Still, sometimes these dream features come together to create a dream with a thematic content that is experienced by a large proportion of the population and, amazingly, that has been described across time, regions, and cultures. If you've ever dreamed of falling, of being inappropriately dressed, or of being unprepared for an exam, then you've experienced our next category of dreams.

TYPICAL DREAMS

Dream themes are called *typical* when many people report having had them at least once. Indeed, people have been offering explanations for these dreams—of being chased, of falling, or of losing one's teeth—for millennia. (According to a dream interpretation dictionary written by the Duke of Zhou in the eleventh century BCE, dreams of your teeth falling out mean that your parents may be greeted by a misfortune.) So it's somewhat surprising that the first major scientific study of typical dreams didn't appear until 1958,[3] when researchers investigated the prevalence of 34 typical dreams in Japanese and American students. Although some cross-cultural differences were noted—Americans reported fewer dreams of fire and more of nudity—the similarities were striking. In both groups, dreams of being attacked or pursued; of falling; of trying again and again to do something; of school, teachers, or studying; and of sexual experiences all figured—with almost identical rank orderings—among the six most frequently reported dream themes. The bottom four typical dreams—dreams of creatures that are part animal and part human, of being buried alive, of seeing oneself in a mirror, and of being hanged by the neck—were also virtually identical in both their frequency and ranking in the two groups. The distribution of the remaining dream themes in the two groups (failing an exam, flying, seeing yourself as dead, or having your teeth fall out) also showed more similarities than differences.

It was another forty years before Tony and fellow Canadian dream researcher Tore Nielsen followed up on these findings. Taking the 1958 study as a starting point, Tony and Tore developed the 55-item *Typical Dreams Questionnaire* and used it to investigate typical dreams in students and sleep-disordered patients.[4] One key finding was how consistent students' prevalence profiles of typical dreams were from one year to the next as well as across student populations from different regions of Canada. Another, even more striking result was how stable people's typical dream profiles had remained over decades. For example, the four most frequently reported dream themes in the 1958 study of American and Japanese students were all among the top five reported by Canadian students some forty years later. What's more, subsequent studies in Germany and Hong Kong have revealed remarkable similarities in the rank order of dream themes across all of these populations.

To give you a better idea of the prevalence of the most common typical dreams, we pooled the findings from several studies that have used the *Typical Dreams Questionnaire* (including those from China and Germany) and compiled a list of the top fifteen typical dreams. The rankings, based on results from 2,000 university students, including 1,500 women and 500 men, are presented in Table 10.1.

Several interesting observations can be made from these findings.

Table 10.1

PREVALENCE OF TOP 15 TYPICAL DREAM THEMES

Rank	Typical Dream *Have you ever dreamed of . . .*	Total Prevalence	Percent of Women	Percent of Men
1	being chased or pursued, but not physically injured	85	86	82
2	sexual experiences	78	75	85
3	school, teachers, studying	77	80	68
4	falling	76	77	75
5	arriving too late (e.g., missing a train)	65	67	59
6	a person now alive as dead	61	65	49
7	being on the verge of falling	59	61	55
8	flying or soaring through the air	56	54	62
9	failing an examination	54	58	46
10	trying again and again to do something	52	51	56
11	being frozen with fright	49	52	43
12	being physically attacked (e.g., beaten, stabbed, raped)	47	48	46
13	a person now dead as alive	44	46	39
14	vividly sensing, but not necessarily seeing or hearing, a presence in the room	43	44	42
15	being a child again	41	42	40

First, no typical dream is truly "universal." Only four of the top fifteen themes have a prevalence exceeding 70 percent and none are higher than 85 percent. Second, themes of being chased or pursued, of sexual experiences, of school, teachers, and studying, and of falling are the most commonly reported typical dreams in both men and women. Third, many of the typical dreams most frequently discussed in popular media aren't all that common. For instance, only 35 percent of people report ever having dreamed of being inappropriately dressed, and fewer than 30 percent report ever having dreamed of being unable to find (or embarrassed about using) a toilet, of their teeth falling out, or of finding money. Fourth, the studies show consistent gender differences. For instance, women are more likely than men to report dreams about school, teachers, and studying, or about a living person who is dead in the dream, or about failing an examination, but they are less likely to report having dreamed of sexual experiences or of finding money.

Finally, although many typical dreams are negatively toned, several of them are positive dreams, including dreams of flying (56 percent), discovering a new room in the house (34 percent), having magical powers other than flying (31 percent), and having superior knowledge or mental ability (31 percent). Some of these typical dreams are, on occasion, negatively toned—a fear of falling while flying or of monsters leaping out of a newly discovered room—but on the whole, they are accompanied by positive emotions.

These numbers, however, reflect lifetime prevalence, the percentage of people who have had such dreams at least once in their lives. They don't tell us how often people have the dreams. Tony evaluated the thematic content of 3,000 randomly selected home dreams from 450 individuals and found that only five of the fifty-five typical dreams— falling, flying, a person now dead being alive, being inappropriately dressed, and being unable to find or use a toilet—occurred in more than 3 percent of the dream reports. Many other themes, like finding money, losing teeth, and failing an examination, occurred in less than a half percent of reports. Other researchers have found similar numbers, including in laboratory-based REM dream reports.[5] Still, when you add up the frequencies of all fifty-five categories of typical dreams, there's a better than fifty-fifty chance that one of these themes will pop up in your dreams tonight.

RECURRENT DREAMS

In the early 1970s, the British neuroscientist Bernard Katz gave a series of lectures at Harvard Medical School, where Bob was working. Katz had won the Nobel Prize in medicine in 1970, and the school amphitheater was packed. At the start of his final lecture, Katz looked out over the audience and said, "I had that exam dream again last night." A collective moan of recognition rose from the audience. Apparently, those dreams never end.

Exam dreams, dreams of going to school in your pajamas (or less), of teeth falling out, of forgotten plane tickets or passports—who needs these repetitive dreams? This might seem like a rhetorical question, but it's not. Certainly, Katz didn't have to worry about not having read the textbook or having forgotten to go to class   two common forms of exam dreams—and your risk of forgetting to get dressed before going to work is, hopefully, vanishingly small. So why do some people continue to experience these dreams? As we'll see later in this chapter, their occurrence may not be all that surprising when viewed through the lens of NEXTUP.

Recurrent dreams not only have the same theme from one time to the next, but they also have the same content. Studies show that about 70 percent of adults report having had at least one recurrent dream during their life, sometimes dating back to childhood. Predictably, recurrent dreams tend to elicit feelings of fascination and puzzlement in those who experience them.

Much of what we know about the content of recurrent dreams— including those collected directly from children and young adolescents—comes from a series of studies conducted by Tony and his colleagues.[6] Setting aside trauma-related nightmares, which can also be repetitive (a topic we'll cover in Chapter 13), about 75 percent of recurrent dreams are reported as being negatively toned; another 10 percent contain a mixture of positive and negative emotions. Even when rated with objective scales like those from the HVC scoring system, negative content categories (such as misfortune, failure, aggressive social interactions) appear about ten times more frequently in recurrent dreams than do their positive equivalents (good fortune, success, friendly social interactions). In the end, only about 10 percent of recurrent dreams contain only positive emotions.

Since the precise content of recurrent dreams is invariably idiosyncratic, trying to classify their thematic content into narrowly defined categories can be challenging, if not impossible. In fact, about one-third of recurrent dreams reported by both children and adults feature such unimaginably wide-ranging—and often inimitably bizarre—content that they simply defy classification. A young woman who took part in one of Tony's studies described such a recurrent dream as follows:

I'm walking along a beach. I know my parents are a short distance behind me. I look out at the ocean and see a giant, pink capital letter A rising out of the water. It says, "I am the letter A, follow me!" Its voice is deep and powerful, like the voice of God in some movies. I enter the water and try to swim toward the giant letter but it keeps receding and the waves keep getting larger. I wake up.

Be that as it may, about 60 percent of recurrent dreams have the dreamer dealing with a challenge or threat, either psychological or physical. If the dreamer is being chased or otherwise confronted, the hostile agents in children's recurrent dreams usually involve fictional or folkloric characters, such as monsters, witches, zombies, and ghoulish creatures; adult recurrent dreams, on the other hand, typically feature human characters, including burglars, strangers, mobs, and shadowy figures.

Nonthreatening content in recurrent dreams takes many forms, such as descriptions of objects or settings, mundane activities, or interactions with people posing no danger to the dreamer. Positively toned recurrent dreams typically involve themes of finding oneself in a bountiful environment, of flying or soaring through the air, of discovering or exploring a secret room, or of excelling at a physical activity such as dancing or playing a sport.

Some of the typical dream themes we just discussed also characterize recurrent dreams, especially in adults. So, in addition to themes of being chased or pursued, people report recurrent dreams of losing their teeth, failing an examination, being unable to find or use a toilet, or driving a vehicle that is out of control. Strangely and somewhat paradoxically, many of these themes are absent from children's recurrent dreams.

In this book, we've mentioned several times that dreams in general—and late-morning dreams in particular—are often tied to people's most salient experiences and emotional concerns, but without directly incorporating the memories of these events into the dream narrative. Recurrent dreams may be one intriguing way the dreaming brain goes about depicting these concerns, often metaphorically, over time. Having previously identified a dream theme—such as being unprepared for an exam or driving a car that's out of control—that embodied some concern and drew a strong response from the dreamer, the dreaming brain returns to this image or metaphor later when a similar experience or concern gets tagged for further processing. We'll have more to say about when and why recurrent dreams occur in Chapter 12. But for now, let's turn our attention to those dreams in which negative emotions become so powerful, they jolt people right out of their sleep.

NIGHTMARES

With their overwhelming emotions and gripping storylines, nightmares have long been a source of fascination. In 1819 John Waller, a surgeon in the Royal Navy of London, remarked that few afflictions were more universal among all classes of society than the nightmare. He was probably right. Nightmares affect a significant proportion of children of all ages and occur at least once a month in 10 to 30 percent of the general adult population. In fact, approximately 85 percent of adults report at least one nightmare per year, and lifetime prevalence of nightmares is close to 100 percent.

There is no universally agreed-upon definition for nightmares, but they are usually described as highly disturbing dreams that seem to awaken the sleeper. The awakening criterion allows researchers to differentiate nightmares from their generally less intense cousins, bad dreams—negatively toned dreams that do not awaken the sleeper. Researchers and clinicians also distinguish between trauma-related nightmares that more or less accurately replay elements from a traumatic event and the more common *idiopathic* nightmares, which occur for no apparent reason.

Many people, and parents of young children in particular, confuse nightmares with sleep terrors, a biologically distinct sleep disorder. You can tell them apart by their distinctive features. Nightmares occur largely during REM sleep in the second half of the night and, upon awakening from a nightmare, people are quickly oriented (awakening fully and realizing that the experience was a dream) and easily recall the nightmare's vivid imagery and detailed storyline, even later that morning. By contrast, sleep terrors occur in deep (N3) sleep, usually during the first hours of sleep. They are accompanied by sudden and intense autonomic activation—your heart rate can double or triple in mere seconds—occasional blood-curdling screams, marked confusion upon awakening, and an absence of dream recall beyond a possible isolated image. Moreover, amnesia for the entire episode is typical upon awakening in the morning.

Until recently, much of what we knew about the content of non-traumatic nightmares was based on either personal interviews (asking people to describe a recent nightmare) or questionnaire data (asking people to select from a list of nightmare themes). But such studies, which have found high prevalence for themes of falling, being chased or assaulted, and of paralysis or death, are probably biased. For one thing, these studies normally ask participants simply to recall and report any nightmare they can remember. Given the fragile nature of our long-term memory for dreams, most people end up reporting a particularly intense, unusual, or otherwise salient nightmare that often dates back years or decades. Nightmares with themes of being chased and of death certainly would fit this description. In addition, nightmares with themes of falling or being paralyzed are most likely due to other commonly experienced parasomnias. Parasomnias are a class of mostly unpleasant experiences or behaviors that often occur on the edges of sleep, such as hypnagogic jerks (sudden twitches or feelings of falling as you're falling asleep) or instances of sleep paralysis, discussed in Chapter 4, that typically occur as you're waking in the morning.

To be sure we're all talking about the same thing, we need to agree on clear definitions of nightmares and bad dreams, and we must eliminate potentially confounding disorders such as sleep terrors. Then, we can ask people to keep a detailed log of all dreams remembered for a period of days or weeks. The process is obviously quite demanding, and most of the dream reports collected won't even be nightmares. But a few years back, Tony and his then doctoral student Geneviève Robert conducted a large, in-depth study looking for bad dreams and nightmares in 10,000 dream reports collected from 572 participants.[7]

Of the 10,000 dreams, about 3 percent were nightmares and another 11 percent qualified as bad dreams. Overall, one out of every seven dreams recalled by people in their natural sleep environment contained strong negative emotions. Moreover, the emotions reported in their disturbing dreams weren't limited to fear; about 35 percent of the nightmares and over half of the bad dreams contained other negative but equally intense primary emotions, including anger, sadness, confusion, and disgust. Not surprisingly, nightmares were more emo-

Table 10.2

THEMATIC CATEGORIES
IN NIGHTMARES AND BAD DREAMS

Theme	Description
Physical aggression	Threat or direct attack to one's physical integrity by another character, including sexual aggression, murder, being kidnapped
Interpersonal conflicts	Conflict-based interactions involving hostility, opposition, insults, humiliation, rejection, infidelity, lying, etc.
Failure or helplessness	Difficulty of the dreamer in attaining a goal, including being late, lost, unable to talk, losing or forgetting something, and making mistakes
Health-related concerns and death	Presence of physical illness, disease, health-related concerns, or death of a character or of the dreamer
Apprehension or worry	Dreamer feeling afraid or worried about someone or something, without an objective threat being present
Being chased	Dreamer is chased by another character but not physically attacked
Evil presence	Seeing or feeling the presence of an evil force, including monsters, aliens, vampires, spirits, creatures, ghosts, etc.
Accidents	The dreamer or another character being involved in an accident, including vehicular crashes, drowning, slipping, falling, etc.
Disaster and calamity	Plausible events ranging from relatively small-scale anomalies such as a fire or flood in one's house or neighborhood to larger-scale disasters such as earthquakes, war, the end of the world, etc.
Insects and vermin	Infestation, bites or stings from insects, snakes, etc.
Environmental abnormality	Bizarre or implausible events appearing in the dream's environment
Others	Idiosyncratic or infrequent themes such as being naked, being in an insalubrious environment, and being unable to find or being embarrassed to use a toilet

tionally intense and bizarre than bad dreams, which in turn were more intense and bizarre than other dreams.

Tony and Geneviève identified a dozen thematic categories for nightmares and bad dreams and then came up with precise definitions that allowed judges to reliably score the dreams. Table 10.2 describes these categories, and their observed frequencies are given in Table 10.3. (The careful reader might note that these combined frequencies add up to 146 percent, not 100 percent. This is because many dreams included more than one of these themes.)

As Table 10.3 indicates, themes involving physical aggression and

Table 10.3

FREQUENCY OF THEMES
IN NIGHTMARES AND BAD DREAMS

Theme	Nightmares (%)	Bad dreams (%)	Combined (%)
Physical aggression	49	21	32
Interpersonal conflicts	21	35	30
Failure or helplessness	16	18	17
Health-related concerns and death	9	14	12
Apprehension or worry	9	13	11
Being chased	11	6	8
Evil presence	11	5	7
Accidents	9	5	6
Disaster and calamity	5	6	5
Insects and vermin	7	4	5
Environmental abnormality	5	4	4
Others	7	10	9

interpersonal conflicts were the most frequently reported, followed by failure or helplessness, health-related concerns or death, and apprehension or worry. Nightmares were significantly more likely than bad dreams to contain themes of physical aggression, being chased, evil forces, and accidents, whereas themes of interpersonal conflicts were significantly more frequent in bad dreams. Of note, several themes commonly reported in questionnaire studies of nightmares, such as feelings of paralysis or suffocation, were entirely absent from the sample of 10,000 dreams; themes of falling occurred so infrequently (1.5 percent of all nightmares and bad dreams) that they were reclassified under accidents. This finding confirmed our suspicion that such themes are probably attributable to parasomnias such as hypnagogic jerks, sleep paralysis on awakening, or sleep apnea, a sleep-related breathing disorder.

Tony and Geneviève also found that men's nightmares were more likely than women's to contain insects or disasters like floods, earthquakes, and wars, while interpersonal conflicts were twice as common in women's nightmares. Finally, the most frequently reported causes for awakening from a nightmare were the presence of an immediate threat (42 percent), the intensity of the emotions experienced (25 percent), and intentionally waking up to escape the nightmare (14 percent).

Taken together, these findings tell us that most disturbing dreams contain threats to survival, security, or self-esteem. They also show us that although nightmares and bad dreams share many characteristics, nightmares—with their more intense emotions, greater bizarreness, and tendency for more violent themes—represent a rarer and more severe expression of the same basic phenomenon.

Sexual Dreams

Given the long-standing clinical and popular interest in sexual dreams and their repeated "top three" standing among typical dreams, this dream category has received surprisingly little scientific attention—which probably says more about dream researchers than dreamers themselves. In 1953 the Kinsey Institute reported that two-thirds of women reported having had overtly sexual dreams at some point in their lives, and that virtually all men did. Approximately 40 percent of women had experienced a sexual dream with orgasm; about 80 percent of men reported nocturnal emissions, with or without accompanying dreams. A little over a decade later, Hall and Van de Castle found that 8 percent of 1,000 dream reports collected from university men and women contained sexual activities, and men reported such dreams more often than women did (12 percent versus 4 percent). Men were also twice as likely as women to report sexual dreams involving unfamiliar partners; women's sexual dreams were 2.5 times more likely to contain a known character, particularly a current partner. It would be another forty years, however, before a handful of other researchers, including some in Tony's lab, would expand on this research.

Numerous questionnaire studies have found that men are more likely than women to report having had a sexual dream. But a more nuanced picture emerges when the frequency of sexual dreams is looked at not with one-off questionnaires but in actual dream diaries. In one study, 287 participants, including both students and non-students, recorded 5,500 dream reports in daily dream diaries.[8] Tony and his colleagues found no significant difference between the prevalence of sexual dreams in the dreams of men (7 percent) and women (6 percent). Moreover, these percentages remain essentially the same in a larger, ongoing study of over 10,000 dream reports from almost 600 men and women. Thus, while a greater proportion of men than women report having experienced at least one sexual dream in their life, these dreams appear with about the same frequency in the everyday dreams of both men and women. Interestingly, the frequency of people's sexual dreams, regardless of gender, correlates not so much with how often they engage in sexual activities as with the time they spend *thinking* about sex.[9]

The observed differences in sexual dream frequency between more recent studies and Hall and Van de Castle's data may be partially due to sample composition (college students versus a mix of student and nonstudent adults). But it's also possible that women actually experience more sexual dreams now than they did fifty years ago and feel more comfortable reporting them, in both cases due to changing social roles and attitudes.

Within these dreams, sexual intercourse is the most frequently reported sexual activity, followed in order by sexual overtures, kissing, fantasies, genital contact, oral sex, and masturbation. Orgasms, however, are reported in fewer than 4 percent of both men's and women's sexual dreams. And as might be expected, women are more likely to describe at least part of the sexual activity as being unwanted.[10]

In line with Hall and Van de Castle's original findings, more recent studies show that while current or past sex partners figure in up to 30 percent of women's sexual dreams, they are present in only 10 to 15 percent of men's sexual dreams. Even in studies of adults in committed relationships, fewer than a third of all reported sexual dreams involve the dreamer's current partner.[11] So who *do* we dream of having sex with? As was the case some fifty years ago, familiar characters, including friends, acquaintances, and public figures, still appear twice as often in women's sexual dreams as in men's, while strangers (including multiple partners) figure in about twice as many of men's sexual dreams. But in one recent study,[12] individuals in romantic relationships of greater duration and reporting higher relationship satisfaction and frequency of sexual activity with their partner were more likely to have erotic dreams involving their current partner, whereas individuals who reported "cheating" on their partner were more likely to have dreams involving an acquaintance or ex-partner. Similarly, another study found that dreams of infidelity were more common in people with past experiences with an unfaithful partner as well as in people with higher levels of romantic jealousy.[13]

And speaking of dreams and jealousy, Tony isn't the only dream researcher to have been told, either by a friend or study participants, a story along these lines: "I was sound asleep when all of a sudden my spouse whacked me across the head! I woke up startled, and when I asked her what the hell that was all about, she told me I'd been cheating on her—in her dream!"

DREAM CONTENT AND NEXTUP

In Chapter 9, we saw how our dreams—from the brief hypnagogic thoughts and images that take place at sleep onset to the more complex and immersive experiences associated with REM sleep—share the same formal features, albeit to varying degrees. Similarly, we have now seen some of the commonalities that exist in the content of dreams, including their overarching themes. But even though two people may report dreams about missing a train, the details of those dreams will invariably differ in some ways, such as how they were planning to get to the train station, what the surroundings looked like, the season and time of day, what made them late, the anticipated consequences, and the roles of any other characters in the dream. Depending on your perspective, the two dreams might seem either very similar or very different from one another.

These similarities and differences raise two important questions. First, why do the same central themes appear across so many of our dreams; and second, where do all the finer details that end up woven into our dreams come from?

Let's start with the second question. We know from a slew of home and laboratory studies that when our brains dream, they usually incorporate events from the day before—or, more rarely, from several days before the dream—what Freud referred to as day-residue. But rather than incorporating an entire episodic memory from the preceding day's events, our brain takes only pieces of them: objects, settings, people, impressions, fleeting thoughts, or bits of conversations. These bits of day-residue get combined with details from older, weakly related memories—some from weeks or months earlier, others from further back in our lifelong collection of autobiographical memories—to provide the details that make up our personal dream narratives. This is exactly what NEXTUP would predict.

But how do we get from these mundane daytime experiences to dreams of failing an exam, of losing teeth, of being chased, or of having superpowers? After all, most of us don't spend our days failing exams, being chased by zombies, having sex with our exes, or flying over gloriously bountiful places. This is where a second, more important memory source of dreams comes into play.

We saw in Chapter 8 that the brain, probably guided by the default mode network (or DMN, which is active in the brain whenever an individual isn't focused on specific tasks), uses downtime as well as the sleep-onset period to identify and tag particularly salient experiences or concerns in need of additional attention so that it may process them later during sleep, including in our dreams. Consistent with this idea, dozens of studies have shown that dreams—particularly late-morning home dreams and laboratory REM dreams—are much more likely to incorporate emotionally salient waking-life experiences and concerns than they are to incorporate less emotional events. For example, a recent study carried out in the laboratory of Mark Blagrove in Swansea, Wales, found that when participants took a few minutes each night to identify and rate the emotional intensity of the main events of their day, it was the more emotionally intense events that provided the elements—a character here, an object there—that were subsequently incorporated into their dreams.[14]

NEXTUP, like many dream theories before it, proposes that dreams are often intimately tied to our most salient experiences and concerns without directly incorporating the memories of these events into the dream narrative. Nor do dreams offer concrete solutions to the problems posed by these experiences and concerns. This is one of the most confusing aspects of dream construction. If a barely avoided car accident is the impetus for a subsequent dream, why is the dream about being on bumper cars at an amusement park? Why doesn't the memory of the actual near-miss event appear in the dream? As we noted earlier, episodic memories, dependent on hippocampal activity, are not reactivated during REM sleep and hence wouldn't be available for incorporation. But then where does the image of the bumper cars come from? How does the brain know that it's relevant to the day's events?

Two of Bob's studies, both previously mentioned, provide one answer. We know from his Tetris study that people can dream about playing Tetris even when both of their hippocampi have been destroyed, for example, by carbon monoxide poisoning. So even without a hippocampus, a trace of this memory must be in the rest of the brain and can be activated, even if it can't be brought back into conscious awareness.

The other clue comes from a study, described in Chapter 5, that Bob did with Jessica Payne. In that study, participants tried to memorize lists of words played on a tape recorder. The participants were not told that each list was made up of words strongly related to some central word, like *doctor*, that was not on the list and hence was never heard. But when participants later tried to recall the words they had heard, a large fraction of them "recalled" hearing the central words, like *doctor*, although they weren't on the list. Even without hearing the word, participants nonetheless associated it with the other words they did hear; so much so, in fact, that they thought they *had* heard that central word.

One detail of the experiment is particularly important here. The lists were created by asking students in an introductory psychology course to write down the first ten words that came to their mind when they thought, for example, of *doctors*. Jessica then built each list from the most commonly reported associated words. The entire protocol is remarkably similar to what NEXTUP proposes happens during dream construction, but with two critical differences. First, when brains dream, instead of some professor asking students to come up with associations, the brain searches through its own networks of associated memories and concepts to find related items. And second, the brain searches for weak, rather than strong, associations. But the experimental protocol is critically similar to the process of dreaming because it specifically excludes the basis for the list—in the case of dreaming, the emotionally salient event or concern identified for nocturnal processing—from the final list of items incorporated into the experiment or dream narrative. And much as the absent words representing the gist of each word list were not explicitly heard by Jessica's subjects in the experiment, so also the events and concerns that lead to the generation of lists of dream elements are not themselves incorporated into the dreams.

Nevertheless, just as Jessica Payne and many others have shown that unheard gist words are (falsely) recalled by their participants, so NEXTUP proposes that the dream elements become associated with the generative waking event, despite its absence from the dream.

Of course, in our dreams, these lists of associated memory elements are not presented as tables or lists. Instead, they are incorporated into embodied dream narratives. The use of narrative is not surprising; we use it in waking as well. Whether in movies or books or simply in storytelling, humans use dramatization, with its similes, figurative plots, and metaphors, to describe and present the emotional events and concerns of our lives. What our brains do while dreaming is not that different from what they do when we go to the theater—they imagine and explore possibilities embedded in a narrative with the hope of gaining new understandings about ourselves and the world we live in.

And just as movies and plays typically involve themes of universal human significance, so our dreams are often centered on common themes—themes of being chased or running late, being unprepared for an exam, soaring through the air, falling ill, dying, or discovering something wondrous. These common dream themes allow the dreaming brain to ask key what-if questions and, more important, to explore possible answers by plunging us into virtual worlds that are consciously perceived and reacted to—worlds that evolve in response to our own thoughts, feelings, and actions within the dream.

And much like our stereotypical preferences for movie genres, the kinds of stories the dreaming brain chooses to tell itself also depend on what thematic contents we as individuals most strongly relate to. This phenomenon is probably why adults are more likely to be confronted by burglars, mobs, and shadowy figures while children are pursued by angry monsters and witches. It may also be why themes involving interpersonal conflicts are generally more frequent in the bad dreams and nightmares of women while themes of disasters, calamities, and war occur more often in those of men. Our dreams are both universal and unique, and they all help us to explore a multitude of worlds that might be.

Chapter 11

DREAMS AND INNER CREATIVITY

A BOOK ABOUT DREAMS MUST INCLUDE A DISCUSSION on creativity. At the simplest level, we can examine the sheer creativity seen in our dreams: the strange and unpredictable juxtaposition of people, places, events, and concepts; the brilliance of visual imagery; and the strange, seemingly metaphorical transformation of our waking thoughts, emotions, and experiences into the landscape and narratives of our dreams. If this description seems to have an almost rhapsodic flavor to it, that's because we are so often struck by these unique features of our own dreams.

In describing the creativity of dreaming in this way, we are not talking about the function of dreams or the brain mechanisms that generate them. We're just describing the truly remarkable nature of the phenomenology of dreaming, the experience of dreaming, and the sense of wonder and delight when we awaken remembering a particularly intriguing dream, vividly recalling and marveling at its content. This is the magic of dreams, and no explanations of mechanisms or functions can diminish it. It is what, for thousands of years, has driven people to wonder about their dreams. On some level it is what drove us to write this book and, we suspect, drives you to read it. But as scientists, the mechanisms and functions of dreaming make this "magic" even more exciting to us, as does the potential usefulness of all these ingenious processes in making people more creative during their waking lives.

Dreaming facilitates our waking creativity in two different ways. One is by directly facilitating problem solving. We've talked about sleep's role in problem solving both in Chapter 5, when we discussed the functions of sleep, and in Chapter 7, when we discussed the function of dreaming. There are lots of famous examples of problems being solved in dreams, though only rarely are the solutions in explicit form.

CREATIVITY AS PROBLEM SOLVING

Three groundbreaking scientific discoveries have been attributed to dreams. In two of these, the solution to a scientific question appeared fully formed in the dream. First, in 1869, the Russian chemist Dmitri Mendeleyev had a dream that he later recounted: "I saw in a dream, a table, where all the elements fell into place as required. Awakening, I immediately wrote it down on a piece of paper."[1] That table later became known as the periodic table of the elements, which chemistry students still study today, 150 years later.

In the second, in 1921, the German Jewish pharmacologist and psychobiologist Otto Loewi dreamed of an experiment that allowed him to demonstrate that nerve cells released chemical neurotransmitters to communicate with one another—an experiment that won him the Nobel Prize in 1936. His dream, and the prize, are probably all that kept Loewi alive after the Nazis arrested him in 1938. His release was secured in return for his "voluntarily" giving all his research to the Germans.

But in the third case, the solution was not explicitly presented in the dream. It was 1865, and the German chemist August Kekulé was struggling to decipher the structure of the chemical benzene, which was known to contain six carbon atoms. Six carbons could be combined in a molecule in many ways, some of which can be seen in Figure 11.1. But Kekulé had shown that all the carbon atoms behaved

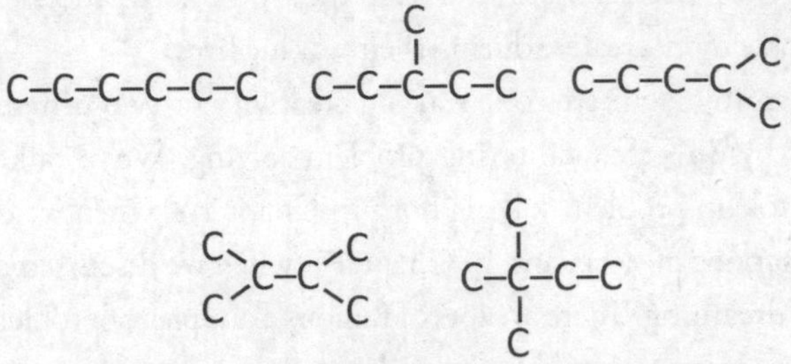

C–C–C–C–C–C   C–C–C–C–C   C–C–C–C⟨C
                             |
                            C

C⟩C–C⟨C   C–C–C–C
                 |
                 C

Figure 11.1. Possible six-carbon configurations. Individual carbon atoms can connect to up to four others.

the same way, which is not the case for any of these structures. Atoms sticking out from the side or at an end would behave differently from those in the middle of these structures.

Then Kekulé had a momentous dream. Here's how he described it in his speech at a jubilee marking the twenty-fifth anniversary of his discovery:

> *I was sitting writing at my textbook but the work did not progress. . . . I turned my chair to the fire and dozed . . . the atoms were gamboling before my eyes. . . . My mental eye, rendered more acute by the repeated visions of this kind, could now distinguish larger structures of manifold conformation: long rows sometimes more closely fitted together all twining and twisting in snake-like motion. But look! What was that? One of the snakes had seized hold of its own tail and the form whirled mockingly before my eyes. As if by a flash of lightning I awoke . . . I spent the rest of the night in working out the consequences of the hypothesis.*[2]

In his dream, Kekulé didn't see the structure of benzene per se. He saw the unlikely image of a snake swallowing its own tail—a previously unexplored possibility. That image led him to another unexplored possibility, namely, that the benzene molecule might form a ring with each of the six carbon atoms attached to two others (Figure

Figure 11.2. The ring structure of benzene.

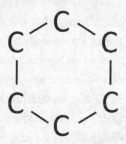

11.2). With that, he had discovered the structure of benzene, the first organic molecule found to have a ring structure.

Examples abound of dreams providing creative insights into inventions (the sewing machine, by Elias Howe), novels (*Frankenstein*, by Mary Shelley), and songs ("Yesterday," by Paul McCartney). Almost every form of creative endeavor has famous examples of dream inspirations, leading some to suggest that creative problem solving is *the* function of dreaming.

But there are problems with this conclusion. First, many of these iconic dreams came from the hypnagogic period, after pondering a problem shortly before sleep. As we discussed earlier, the physiology of the hypnagogic state, its dream content, and the function of its dreams are all unique to that state. Its physiology is more directly linked to prior waking physiology, and its dreams reflect current waking concerns more strongly than later sleep stages. Indeed, the hypnagogic period has been exploited precisely for its ability to reflect creatively on such concerns.

Thomas Edison, whose inventions led to over a thousand patents, developed just such a technique. Bob learned of it when touring Edison's former research laboratory in Fort Myers, Florida. On the tour he was shown an armchair with a tin plate on the floor, placed in front and to the side of the chair. Here, he was told, Edison would solve problems that were blocking the completion of his most recalcitrant inventions. Edison sat in the chair, arms on the armrests, and held a metal spoon between thumb and forefinger of one hand, over the plate. Then, while pondering the obstinate problem, he allowed himself to slowly drift off to sleep. As he entered hypnagogic sleep, the muscles in his thumb relaxed; the spoon then fell, hit the tin plate, and woke him up—with the solution to his problem clear in his mind!

Similar stories can be found about Einstein. But it was Salvador Dalí who provided a detailed report of his use of the technique. Dalí vividly described his "slumber with a key" technique in his 1948 book, *50 Secrets of Magic Craftsmanship*:

> *Seat yourself in a bony armchair, preferably of Spanish style, with your head tilted back and resting on the stretched leather back. Your two hands must hang beyond the arms of the chair, to which your own must be soldered in a supineness of complete relaxation.... In this posture, you must hold a heavy key which you will keep suspended, delicately pressed between the extremities of the thumb and forefinger of your left hand. Under the key you will previously have placed a plate upside down on the floor. Having made these preparations, you will have merely to let yourself be progressively invaded by a serene afternoon sleep.... The moment the key drops from your fingers, you may be sure that the noise of its fall on the upside down plate will awaken you.... [This] is exactly, and neither more nor less, what you needed before undertaking your virtuous afternoon labors.*[3]

A second reason we reject such problem solving as the function of dreaming is that, for all the brilliance demonstrated in these classic examples, creative breakthroughs in dreams are rare. Few of us can point to a major—or even minor—creative breakthrough that was revealed to us in a dream, and almost none can point to one that took place in the past week. And even when insights about or solutions to problems are derived from a dream, they typically only crystallize into their full-fledged form when thinking about the dream *after awakening*, not during the dream itself.

Finally, these dreams of discovery don't happen to just anyone at just any time. Each of the dream-related scientific discoveries we described came after the scientists had spent months, if not years, obsessing over a problem and working relentlessly to resolve it. Mendeleyev spent years working on various drafts of the periodic table of the elements before settling on the one that appeared to him in his dream. So, although "problem solving" dreams clearly do occur, they are quite rare and cannot reflect the normal modus operandi of dreaming.

It is a separate question whether techniques such as those of Edison and Dalí can hijack the normal behavior of hypnagogic dreaming for our own creative advantage. Partly to answer this question, Adam Horowitz, a graduate student at MIT being advised by Bob, is developing Dormio™, a modern-day electronic version of this technique. It prompts users with a prerecorded message of their own design (for example, "Think about a fork") as they're falling asleep. It then detects their sleep onset, awakens them, records a hypnagogic dream report, repeats the recorded message, and does it all again. Here's an actual transcript from one subject. The text in italics was spoken by Dormio.

Q1: *You're falling asleep. Think of a fork, a fork.* [A pause as participant falls asleep; then Dormio wakes them up] *Tell me, what are you thinking?*

A1: A fork in a supermarket; and I'm trying to use it to cook burgers on the seaside; and my friends are there; and it's very comfortable and it's a metal fork. One fork.

Q1b: *Can you tell me more?*

A1b: The house . . . It's a house I've been in. And a sink, and there are forks in the drawers but there's only one fork I'm using. And there's a charcoal grill; it's making beef burgers and we bought the forks at the supermarket and there's a lot of smoke.

Q2: *Okay; you can start falling asleep again.* [Pause until sleep onset is detected again] *Tell me, what are you thinking about?*

A2:   A speech about a fork. A monkey is holding it. An eagle is carrying it across the trees. It's a wooden fork. And the family is happy to see the fork. And they're putting it in a pumpkin. And the secret agent is using the fork to go into the headquarters. People in sunglasses are taking the agent out. And the agent throws the fork into a tree.

Q2b: *Can you tell me more?*

A2b: The fork is on the ground. And a child picks it up and throws it to a bird. And the bird lays an egg with a fork . . . in it. And there's a caterpillar.

Q3:   *Okay; you can start falling asleep again. Remember to think of a fork.* [Pause until next detected sleep onset] *Tell me, what are you thinking?*

A3:   A fork the size of a lightbulb. And there was a city in the lightbulb. The fork is found under a sink. There is a maze in the fork. And the water from the sink is flowing through the maze. And it gets filled up.

Q3b: *Can you tell me more?*

A3b: A fork comes in plastic tubes in a waterfall. And it flows into space. And they're on the moon, and the fork is planted as a flag. The sun grows bigger. The maze is made of grass and shaped like a brain. There are cats in the maze, and spiders, and the spiders—their legs are made of metal and red and brown. And they have ears, and someone scratches the ears with a fork. And the fork is used to scoop ice cream into a Styrofoam cup.

The progression of these reports is reminiscent of what we described in Chapter 8 when talking about Erin Wamsley's study of the *Alpine Racer II* videogame. Reports A1 and A1b could have been descriptions of activities in a person's day. They describe normal events. Report A2, while describing physically possible events, has become quite bizarre, and Report A2b becomes frankly impossible. By Reports A3 and A3b, the descriptions have become not only bizarre and impossible, but hard to imagine. But they are all about forks.

Do these Dormio reports creatively solve any problems? It doesn't seem likely, although there weren't any fork-related problems that needed solving. Do they seem to be finding weak associations that are creatively combined? It's not obvious, but Bob was struck by the image in Report A2 of the fork being put into a pumpkin and then being used by a secret agent. It reminded him of hearings before the infamous House Un-American Activities Committee back in 1948, when Whittaker Chambers claimed that he had hidden secret microfilms about an alleged Communist agent, Alger Hiss, in a hollowed-out pumpkin. Alas, when Horowitz checked back with the study participant, he had no knowledge of the secret agent Alger Hiss or his pumpkin. The participant suggested that the linkage might have been "pumpkin → orange → agent orange → secret agent." But we'll never know.

Be this as it may, Horowitz's goal is to turn Dormio into an inexpensive consumer product that anyone can use to harness the creative power of their hypnagogic dreams. Bob's goal is to get some good, hard, scientific evidence that will tell us whether the technique can be used to harness our creativity.

The frequency of problem solving in dreams from later parts of the night is less clear. At least some of the breakthrough dreams we described earlier—Otto Loewi's in particular—were clearly not from the start of the night. Unfortunately, no dreams like this have been recorded in the sleep lab, and so we know nothing about what stages of sleep they arise from. But how common are such dreams? Deirdre Barrett, a dream researcher at Harvard University, has chronicled a small number of studies on this question and found that up to a third of participants appear to be able to solve a problem with personal meaning over the course of a week, although only 1 percent of participants trying to dream of a solution to a brainteaser succeeded in doing so.[4]

Having read all this, it's important for you to keep in mind that we have been talking only about dreams *that people recall when they wake up.* We know that the vast majority of our dreams are lost from conscious awareness before we awaken, so it's not known whether these dreams are also producing solutions to our problems. What's more, we talked in Chapter 5 about studies showing that creative solutions to problems can indeed be found at a relatively high rate after a night of sleep, when compared to an equal period of time spent awake. But it's still not known whether these nocturnal solutions arose during unremembered dreams or through completely nonconscious processes.

CREATIVITY AND DIVERGENT THINKING IN DREAMS

We started this chapter by claiming that dreaming fosters waking creativity in two ways, and we have spent the first part examining how dreaming can facilitate waking creativity by enhancing problem solving. But from the perspective of NEXTUP, these instances of explicit problem solving are just the tip of the iceberg. The true creativity found in dreaming is in the creative exploration of associative neural networks, in which our brain ferrets out normally weak associations that are of potential value in addressing these problems. Whether this counts as true creativity is a matter of definition. Robert Franken, in his book *Human Motivation*, defines creativity as "the tendency to generate or recognize ideas, alternatives, or possibilities that may be useful in solving problems, communicating with others, and entertaining ourselves and others."[5]

On the other hand, Mihaly Csikszentmihalyi, at the University of Chicago, takes a much more goal-oriented approach, defining creativity as "any act, idea, or product that changes an existing domain [of knowledge (like mathematics)], or that transforms an existing domain into a new one."[6] He concludes that what counts "is whether the novelty . . . is accepted for inclusion in the domain."[7] This definition implies that the product of the creative act must be of some universal value or significance—a demanding constraint, indeed.

Although the iconic dreams of Kekulé, Loewi, and Mendeleyev described earlier clearly meet this more stringent definition of creativity, our common everyday (or every night) dreams clearly don't. But they do seem to meet Franken's more modest definition. In fact, his phrase "the tendency to generate or recognize ideas, alternatives, or possibilities that may be useful in solving problems" almost exactly matches our description of NEXTUP as "network exploration to understand possibilities." More specifically, this matches the description of REM sleep dreams, which are characterized in Chapter 8 as an almost willful intent to seek out creative associations.

HARNESSING THE CREATIVITY OF OUR DREAMS

As the role of sleep in memory processing becomes more and more apparent, interest in manipulating sleep for our own benefit has grown dramatically. Recently, noninvasive electric, magnetic, and auditory stimulation of the brain—and in some cases, specific regions in the brain—has been used in attempts to enhance sleep-dependent memory evolution. But other techniques have been developed to enhance the benefits of dreaming. Drugs have been identified that enhance lucid dreaming, and visual stimulation through the eyelids has been used to train people to dream lucidly (see Chapter 14). Techniques of collecting sleep-onset hypnagogic dream reports, developed by Edison, Dalí, and Adam Horowitz, have been used to enhance creativity and problem solving.

Some of these dream techniques are not new—Edison's goes back a hundred years—but a massive uptick in research and technological development has taken place in the past ten years. Just last year, Bob went to a dream-engineering symposium[8] at MIT's Media Lab— a self-described antidisciplinary research lab—where over a dozen researchers presented new technologies for enhancing the study and manipulation of dreams. Horowitz's Dormio was just one of them. But in addition to these relatively new methods, there are also older and simpler techniques based on the practice of "dream incubation" that can foster the emergence of creativity and provide solutions to everyday, real-life issues through our dreams.

The practice of dream incubation—various techniques employed during wakefulness to help a person dream about a specific topic or obtain solutions to specific problems—can be traced back four thousand years to the Mesopotamians. But it was fifteen hundred years later, in ancient Greece, that dream incubation became widespread. People seeking healing dreams traveled to shrines dedicated to Asclepius, son of Apollo and the Greek god of medicine, in the hopes of receiving dreams that would reveal the cause of their illness or, better still, a cure for it. And, in a practice that was probably as alarming then as it would be today, nonvenomous snakes were sometimes placed inside the temples and allowed to crawl about among the dream-seeking supplicants. This was not done to induce paralyzing insomnia or terrifying nightmares, but to help induce the desired dreams. Snakes were the emblem of Asclepius, and you can still see the association between snakes and medicine today in the logos of innumerable medical organizations that feature a snake-entwined rod, the staff of Asclepius (Figure 11.3).

Today, people practice various forms of dream incubation to discover solutions to everyday problems and concerns—practical, interpersonal, and emotional. There may be a lack of controlled scientific experiments to support these techniques, but there is a wealth of

Figure 11.3. The Star of Life insignia is displayed on emergency medical services such as ambulances around the world.

descriptive reports of insights attributed to dream incubation. These, combined with what we know about the impact of sleep and dreams on learning, memory evolution, and the exploration of associative memories, lead us to recommend this absolutely free, time-effective, and harmless way of harnessing dream creativity. Dream incubation may even help you discover solutions to your real-life problems.

Methods of dream incubation can be found in dozens of books and online sources, and they run from relatively banal procedures to highly intricate pre-sleep rituals. We have opted to present a simple, step-by-step approach that Tony has been recommending for years to help elicit problem solving dreams. (Whether to include slithering snakes as part of the process is up to you.)

DREAM INCUBATION TECHNIQUE

1. Choose a night when you are not overly tired or under the influence of substances that may negatively affect your sleep, such as alcohol, sleeping pills, or recreational drugs.
2. Take a few minutes to think about the problem you want to target in your dreams. You may find it helpful to ask yourself some questions like these: Am I ready to act on

this problem? How do I feel about the situation at this very moment? What would be different if the problem was resolved?

3. Summarize the problem in a short phrase, question, or one-line sentence. Don't be afraid to change the wording until you find the version that feels right. Write down this incubation phrase and keep it by your bed.

4. When you are ready to go to sleep, tell yourself that you will dream about the problem. Make sure you have a pen and paper or recorder (a smartphone will do) by your bedside.

5. Repeat your incubation sentence to yourself as you fall asleep. If you catch your mind wandering, let those thoughts go and bring your attention back to your phrase.

6. Sleep!

7. Upon awakening, either in the middle of the night or in the morning, lie quietly in bed with your eyes closed. If awakened by an alarm, turn it off and close your eyes again. Give yourself a few minutes to remember as much of your dreams as possible. Only then, open your eyes and write down or record everything you remembered, even if it's just an isolated image or fragment of a dream. Avoid making any judgments about the dream at this point; focus on getting everything down before it's forgotten.

8. Examine how your recalled dreams may relate to your incubation phrase. (It might be helpful to use some of the ideas and techniques described in the next chapter.)

Most important, don't get discouraged if you don't recall a dream or see no link between your dreams and the problem you've targeted. As with most things in life, practice helps. It might take several nights to get to an answer. And if your dream did help address your question, then do it again the next time you're stuck on a problem.

Why and to what extent dream-incubation techniques work remains a matter of debate. Back in Chapter 8, when first detailing the workings of NEXTUP, we proposed that the brain uses the sleep-onset period to tag current concerns, or incomplete processes, for later processing during sleep. We think the dream-incubation method just outlined here—as well as similar techniques—simply helps the brain tag the targeted problem. Then, depending on the networks explored by NEXTUP over the night, and on what dreams you recall upon awakening, you might find a creative insight or solution to the problem. What to do with your slithering helpers, however, we'll leave for you to figure out.

Chapter 12

WORKING WITH DREAMS

IDEAS, METHODS, AND CAVEATS

AS DREAM RESEARCHERS, WE OFTEN HAVE STUDENTS, friends, or strangers tell us one of their dreams, followed by the inevitable "What do you think it means?" (Bob always answers, "You are one sick cookie!") So, let's start by getting one thing out of the way: We don't know what your dreams mean and, we believe, neither does anybody else. By this, we don't mean to suggest that dreams are devoid of substance—this book details many ways that dreams can be viewed as psychologically meaningful creations of the brain. But we do take issue with the notion that behind every dream lies one "true" meaning, and that this singular interpretation can be deciphered by specially trained or gifted individuals.

In Chapter 7, we encountered the notion that dreams serve to carry symbolic or disguised messages as one of the oldest views about why we dream. In fact, the desire to make sense of our dreams— to understand what they *mean*—has probably been around since our ancestors first started remembering their ethereal nightly reveries. But we also saw that dreaming could not have evolved as a mechanism solely destined for interpretation. One reason is that we remember far too little of our nightly dreaming for this idea to make sense. A second reason is that few—if any—of the dreams we *do* remember ever get interpreted, let alone by a so-called specialist. And this last point goes double for your dog's dreams. This is why it's important to differentiate between the biological or adaptive function of dreams—this takes place "live," while the dream is unfolding—and the uses we *choose* to make of the dreams we do remember after awakening, which may include interpretation, creativity, or mere amusement.

How should people interested in using dreams for self-exploration go about doing this? We'll present some options later in the chapter, but let's start with how *not* to go about it.

When an artist creates a painting, sculpture, or poem, she doesn't usually present the work to other people and ask, "Can you please tell me what this means?" Even if she did, the answers would vary considerably from one person to the next, and many would fail to resonate with the artist. Yet this is exactly what many people do with their dreams. Others turn to "dream dictionaries," literally looking up the dream's meaning in the same way they would look up the meaning of a foreign expression or word. But these approaches exclude the dreamer from the interpretative process, which is a problem. To be effective, dream exploration requires the active involvement of the person who experienced the dream. What's more, these "X means Y" assertions are based on the assumption that dreams hold universal meanings— an idea for which there is precious little clinical, and certainly no scientific, evidence.

We believe that dreams, much like art, can hold multiple meanings. There isn't one right way of working with dreams, any more than there is one correct way of appreciating art. But some approaches to dream exploration are more accessible than others; they have been the subject of careful studies and have been found to yield personal insights, whether used in therapy or in everyday self-exploration. These ideas and methods are the focus of this chapter.

CLINICAL USE OF DREAMS

Although it is a potentially helpful tool in achieving greater self-understanding, dream interpretation—or dreamwork, the term preferred by those who view dream exploration as a collaborative effort between therapist and client, is used only occasionally in most forms of psychotherapy. (Use of the term *dreamwork* as presented here should not be confused with Freud's concept of dreamwork described in Chapter 3 and referring to unconscious processes responsible for distorting unacceptable wishes into acceptable dream images.) But in reality many people want and even need to interpret their dreams, be it in therapy or some other setting. (We discussed where this feeling might come from in Chapter 8 in the section starting on page 116.)

This is why clients sometimes initiate dreamwork by recounting a dream to their therapist, even if the topic of dreams has never been broached in previous sessions. Clients quite often share a dream with their therapist in hopes that she will be able to shed light on the dream's meaning or somehow use the material to gain insight into their daily functioning and well-being. Many clinicians, however, do not feel competent to work with clients' dreams (few clinical programs include courses on sleep, let alone on dreamwork), or they believe that dreams are a meaningless, unscientific object of study. In such cases, the client likely comes away disappointed while the clinician misses out on a unique therapeutic opportunity.

Therapists choosing to integrate dreamwork into their practice must decide how they want to do this. For some, the idea of dream interpretation is inherently tied to Freudian dream theory and the need to retrace the dream back to unconscious conflicts and desires. For others, working with dreams implies mastering Jungian ideas of how dreams are structured by myths, archetypes, and both personal and collective levels of the unconscious. In Gestalt-based approaches, different dream elements are understood as projections of both accepted and disowned aspects of the dreamer's personality. Other approaches focus on the emotions experienced during dreams, on interpersonal relations depicted in dreams, on bodily sensations experienced while retelling a dream, and so on. Dozens of techniques and schools of thought are aimed at using dreams for greater self-understanding, each with distinct clinical and theoretical underpinnings.

Over the past few decades, however, interest has been growing in contemporary models of dreamwork that integrate ideas and methods from multiple schools of thought. One such approach is the cognitive-experiential dream model developed by Clara Hill,[1] a psychotherapy researcher and professor of psychology at the University of Maryland–College Park. For many reasons, we have chosen to highlight Hill's model. First, this model has been developed through years of practice, teaching, and research. Second, it brings together many of the most interesting and time-tested ideas associated with specific schools of thought (for example, Gestalt, Jungian, phenomenological, and psychoanalytic). This work adds to the model's richness and also makes it accessible to therapists from different schools of psychotherapy. Finally, the cognitive-experiential dream model has been the subject of considerable research, and a strong body of evidence supports its effectiveness in a variety of settings.[2] Here's an overview of how this model works.

Using the Cognitive-Experiential Dream Model to Gain Insights from Your Dreams

The Hill model is comprised of three stages: exploration, insight, and action. The exploration stage has four steps—description, reexperiencing, association, and waking-life triggers (easily remembered by its acronym, DRAW). In this stage, you are asked to recount your dream in the first-person present tense (as if you were currently experiencing it). You then choose a few salient images from the dream. Next you and the therapist explore these images one at a time, and in sequence:

First you (1) describe the image in as much detail as possible, then you (2) reexperience the feelings associated with the image, (3) provide associations to the image, and (4) identify potential waking-life triggers for the image.

In the next, insight stage, the therapist helps you create a meaning for the whole dream by integrating what was learned during the exploration stage. This collaborative work can take place at several different levels. First, the dream experience itself can be explored as is, in the here and now, without considering its possible relation to anything else. Second, the dream can be examined in terms of its connection to your current life situation, including recent experiences, ongoing concerns, and emotions. Third, the dream can be explored as a function of more complex psychological dynamics, including as projections of the self, in relation to childhood conflicts, or arising from spiritual and existential concerns.

Finally, in the action stage, you consider possible changes in your life based on your understanding of the dream. One common way of introducing this stage is to ask how you would like to change the dream. This question serves as a starting point to consider possible changes that could be brought about in your life. With the therapist's help, you can see how such changes can be achieved.

As with other contemporary models of dreamwork, Hill's cognitive-experiential approach is less about uncovering *the* concealed meaning of a dream, and more about exploring possible personal meanings of a dream in a way that helps you discover something new about yourself, leading to further insight, increased self-awareness, or stronger involvement in therapy.

It can be difficult to quantify the benefits of dreamwork within psychotherapy, but clinicians and researchers have developed a variety of tools and methods to do just that. Here's what they have found.

One of the best-documented benefits of contemporary dreamwork, including Hill's model, is the development of personal insight.[3] This might include insights into how you view yourself, think about your problems, relate to others, or remain affected by distant events.

Studies on the effectiveness of Hill's model find that clients give high ratings to the cognitive-experiential dreamwork sessions. Researchers saw that this model contributed to the client-therapist working alliance, increased client involvement in therapy, and enhanced the understanding of client dynamics and clinical progress. It could also improve clients' well-being, perhaps reducing symptoms of anxiety and depression. That's quite an impressive list for a largely misunderstood and frequently overlooked therapeutic tool.

Personal Dreamwork

When dreamwork takes place in a clinical setting, a therapist is always present to help you explore and understand your dreams. But what if you want to engage in dreamwork on your own? One option is to adapt some of the ideas and techniques found in Hill's model and in other therapy-based models of dreamwork for personal use. Another option, based on the idea that two—or four, or six—heads are better than one, is to work with a small group of people interested in understanding and appreciating their dreams as well as those of others. We'll consider both of these options.

Individual Dreamwork

The first step in working with your dreams is to keep a record of the dreams you remember. In the previous chapter, we suggested methods for eliciting problem solving dreams; these ideas also apply to keeping a dream journal. Specifically, keep a pen and paper or a recorder by your bedside. When you are ready to go to sleep, tell yourself that you will remember your dreams in the morning and, most important, give yourself a chance to remember them. (Bob tells his students, "Repeat, 'I will remember my dreams' three times before going to bed. It's embarrassing, but it works.") When you wake up next morning, don't open your eyes, and don't start thinking about your day. If you were awakened by an alarm, turn it off and close your eyes again. Lie quietly in bed and try to float back into a dreamlike state. Give yourself a couple of minutes to remember as much of your dreams as possible. If nothing comes to you, try slowly changing positions (turning onto your side or back).

When you do remember a dream, keep your eyes closed and review everything you can remember about it. Only then should you write down or record your dream. The important thing is to get everything down before it's forgotten, even if it's just a vague sensation or an isolated image. Once you're done, give your dream a short title. This will help you navigate your journal as the number of entries grows.

Once you've recorded a dream that you want to explore, start by rereading it carefully. Close your eyes and take the time to reexperience the dream—its images, thoughts, and emotions—from start to finish. Next, ask yourself some questions designed to help explore what the dream may mean for you. The following list gives some example questions. The list isn't exhaustive, nor is it definitive, but it's a good place to start.

+ How did you feel in the dream? What was its central emotion? When was the first or most recent time you felt this way?
+ Think of the dream's setting. What was it like being there? Does it remind you of anything?
+ Think of the people in your dream. What were they doing? If you recognize a dream character from waking life, what are they usually like? Who or what else do they remind you of? Can you see parts—liked or disliked—of yourself in these dream characters? How do you feel about these parts?

+ If there was an animal in your dream, what was it doing? How did it make you feel? How would you describe its personality or chief distinguishing feature?
+ What were the main images in the dream? What kinds of associations come to mind when you think about these images now? Can you identify waking-life sources for them?
+ What was on your mind when you went to bed that night?
+ Given your answers to these questions, does your dream— or specific elements of it—remind you of particular situations, experiences, or ongoing concerns in your life?
+ Taken as a whole, what do the answers to these questions suggest about who you are and who you want to become, and how you view and interact with the world around you?

Working with these kinds of questions can help you better understand your dreams, although you will probably need to work with several dreams before becoming comfortable with the process. Moreover, it's important to keep in mind that new information, connections, or insights about yourself or your life circumstances may not arise from the whole dream, but from one of its elements. In other words, don't expect to "understand" your dreams in their entirety, especially when you're starting out or working with dreams that are particularly long, complex, or bizarre. Gleaning information from a central image, theme, emotion, or interaction from the dream is a much more likely, and equally valuable, outcome. Give it a try.

Group Dreamwork

Probably the best-known and most widely used approach to "dream groups" is the *dream appreciation* method developed by the late Montague Ullman, a New York–trained psychiatrist and psychoanalyst who, in the 1960s, founded the Maimonides Dream Laboratory in

Brooklyn. His dream appreciation method,[4] which emphasizes discovery and safety, contains several stages, the most important of which are summarized here.

As the dream presenter, you begin by describing one of your dreams to the rest of the group, who may then ask questions to clarify the dream's content. You then listen to the group members as they each work with your dream as if it were their own (the "If it were my dream" technique). Group members typically begin by discussing the feelings they would have experienced if the dream had been their own and then share personal associations and projections about the dream and what it could mean given their own life circumstances. You then respond to what has been shared by the group and explore how the dream relates to your own life, with a focus on recently experienced events and concerns. The dream is then read back to you, in the second person, after which all members of the group engage in an interactive discussion. At a later session, you may come back to this dream and share any additional insights or thoughts you have had about the dream.

Throughout this process, you are made to feel safe about sharing your dream and any personal and potentially intimate details about your life. You should never feel obligated to share a dream or reveal details about yourself that you don't want to, and you always have the option of stopping the discussion of your dream at any time. Finally, as with the cognitive-experiential model of dreams, other people's own associations with and interpretations of your dream should never be imposed on you.

In addition to a vast descriptive literature on the uses and many adaptations of Ullman's dream appreciation method, studies support the idea that, when used as prescribed, this method can result in many dream-related benefits for the group members, not the least of which are gains in personal insights.[5]

Finding Meaning in a Series of Dreams

Most people interested in dreamwork focus on individual dreams. But working with a series of dreams can also be beneficial. In Chapter 9 we mentioned that, in the 1950s, Calvin Hall became fascinated by how the patterns of dream content seen in a long series of dreams could be used to infer aspects of the dreamer's personality, core conflicts, and concerns. Hall and others have shown that it is possible to extract psychologically meaningful information from a series of dreams reported by an individual unknown to them.

You can use several of their ideas and methods to study your own dreams. In the process, you will likely learn things about yourself that could not be inferred from any single dream. If you examine a series of, say, 25 or 50 of your dreams, you will probably notice patterns in their contents. These might include where your dreams tend to take place; their overall themes; which people most frequently appear in your dreams; how you interact with them, and they with you; how you feel and act in certain settings or in the presence of specific people or objects; and how and when your dreams shift in terms of action, bizarreness, or emotions. If you have a dream diary, or plan on starting one, working with a series of dreams can be a useful—and often overlooked—way of getting the most out of your diary.

Similarly, working with recurrent dreams can also lead to personal insights or better self-understanding. According to several studies, recurrent dreams—and their cessation—may be tied to how well we are dealing with emotional concerns and challenges in our everyday lives.[6] If you have a recurrent dream and wonder why it pops up when it does, pay attention to the emotionally salient events or stressors that took place immediately before each recurrence of the dream. You may find, for instance, that the dream appears after you've been disappointed by someone close to you, or perhaps when you are going through periods of self-doubt, or you are feeling ashamed. And lest you forget, one reason recurrent dreams remain largely unchanged from one occurrence to the next is that you, within the dream, react to their unfolding content with the same thoughts, behaviors, and emotions each time.

By revisiting previously explored possibilities and tracking the dream self's reactions to them over time, your dreaming brain may be able to better gauge your progress—or its lack—in recognizing or resolving key emotional preoccupations.

NEXTUP, Insights, and Waking Sources of Dreams

In most approaches to dreamwork, including those described in this chapter, people are inevitably asked to examine the content of their dream in relation to their waking-life circumstances, or to identify potential waking-life triggers for key elements within the dream. Most people view this task as relatively straightforward, but linking dream content to waking-life sources with any confidence is quite complicated.

Back in Chapters 8 and 10, we explored the question of where the details that are woven into our dreams come from. We saw that determining the waking sources for the different elements found in our dreams is not only tricky, but sometimes impossible. Indeed, studies have found that only about 20 percent of dream material can be confidently traced to waking-life sources, even when dreams are explored in group settings (for example, with Ullman's dream appreciation approach). Given that NEXTUP specifically incorporates novel and unusual associations into our dreams—especially during late-morning periods of REM sleep—this finding shouldn't come as a surprise. But it *can* be surprising, even to dream researchers.

A few years back, Tony was giving a talk to a group of students when something about the feel of the auditorium made him realize that he was dreaming. After scanning the room, he closed his eyes (in the dream) and imagined himself standing on a wide, sandy beach with the ocean surf crashing onto the shoreline. When he opened his eyes, he was indeed standing on a beach, and there were large, rolling waves like he had imagined. But there was something very strange in his dream: There were penguins—*thousands* of penguins. They were all over the place, including a dozen or so standing by his feet. They were just minding their own business, wobbling about, a bobbing mass of black and white, extending into the distance. Tony woke up perplexed. Why had his brain placed a sea of penguins on his otherwise idyllic beach? Why not seagulls, or Frisbees, or better still, nothing at all? He couldn't even remember the last time he had thought about penguins. Try as he might, Tony was unable to find a waking, associative, or even metaphorical source for the birds that made any sense in the context of his current life.

Several days later, while Tony was riding in a friend's car, a billboard caught his eye. It featured two beaches, side by side, each extending into the ocean. The beach on the left had people enjoying the sand and wading into the water. But the one on the right was covered in penguins! It was a tourism ad for some Canadian beaches with children playing on the water, making the point that they were much, much warmer than those beaches farther south in Maine (with all those penguins). Tony had no memory of ever seeing the ad, but he had driven that stretch of road recently. And in all likelihood, his brain had registered the image without his mind ever noticing it. So, when Tony became lucid in his dream and tried to imagine an idyllic beach, his brain searched his memory networks for weak and potentially useful associations, and it came up with an unusual what-if—a throng of penguins, just like in the billboard. Or maybe.... What about those students in Tony's dream who were attending his lecture before he became lucid? Were they, like the penguins, "just minding their own business, wobbling about, a bobbing mass of black and white, extending into the distance?" Was NEXTUP exploring something useful about Tony's students? We'll never know.

When brains dream, they can search for weak, novel associations anywhere, including in events people took little or no note of. This is one reason that linking the contents of our dreams to recent or distant life events is often trickier than people imagine.

In Chapter 8, we saw how Bob's dreaming brain probably strengthened its exploratory associations between his daughter Jessie and the dog lab because of Bob's reaction in the dream and how these associations affected its unfolding. Tony, however, reacted to the beach-penguin-ocean association with mere puzzlement before waking up. His brain likely understood that these associations were not worth strengthening; they appeared to be useless for better understanding of future possibilities. Not everything in a dream has earth-shattering revelations buried in it.

Sometimes, however, the waking source for a dream is as plain as day—or so it may seem. One of Tony's friends (we'll call him Eric) once told him about a dream in which he was at the wheel of his car, stuck in a giant snowbank. Linda, his girlfriend, was sitting beside him. Eric had apparently accidentally driven the car into the snowbank—although he did not remember this part of the dream—and now Linda was mad at him and imploring him to do something. Eric tried turning the wheels left and right, tried to rock the car free by accelerating forward and backward, and even got out of the car to clear the tires. But nothing worked—the snow was too deep. To Eric, the source of the dream was evident. The day before the dream, there had been a snowstorm in Montreal and he'd seen a car in a ditch by the road, its driver working hard to dig it out of the snow. Tony, however, was struck by the feelings Eric said he'd experienced during the dream. Eric hadn't been worried about the accident, the car, or being late for something. Instead, he was upset at his girlfriend for blaming him for their predicament.

Following a few probes and exchanges, Eric saw a new source for his dream: His relationship with Linda. It turns out that Eric and his girlfriend were going through a tough patch and they both felt like their relationship was stagnating. What's more, Eric felt like he was the only one taking concrete actions to improve things while—in his mind—also shouldering much of the blame for their problems. In other words, they were stuck, and the dream was a perfect metaphor for a situation that had nothing to do with actual cars or snowstorms.

Although these examples illustrate some of the uncertainties involved in making sense of our dreams, we believe that the kind of dreamwork described in this chapter can yield personal insights, and that it makes sense from the perspective of NEXTUP. We've previously seen that dreaming plays an important role in how memories selected for processing evolve across the night. We've also seen that when brains dream, they identify and strengthen associations that in one way or another embody salient experiences and emotional concerns; furthermore, the brain calculates that the associations may be of use in resolving these or similar concerns, either now or in the future. Finally, we saw that one fascinating aspect of NEXTUP is its simultaneous monitoring of how the dreamer reacts to the dream world and, in turn, how the dream world reacts to the dreamer's own thoughts, actions, and emotions in the dream, thereby producing the ever-evolving dream narrative.

By reflecting on key features of a dream, such as its setting, characters, main images, and theme, you are, in a very real sense, reflecting on the experiences and concerns embodied in this dream world. And by examining the thoughts and emotions you experienced during the dream, you're not just delving into how these concerns likely make you feel and behave in waking life. You're also peering into how your ongoing thoughts, actions, and emotions during the dream influence your brain's choice of associations and possibilities during its interactive construction of the dream. If this process of exploration doesn't offer us a chance to learn something about ourselves and the world we live in, we don't know what does.

Chapter 13

THINGS THAT GO BUMP IN THE NIGHT

PTSD, Nightmares, and Other Dream-Related Disorders

OVER THE COURSE OF THIS BOOK, WE HAVE MEN-
tioned and described features of a number of dream-related disorders,
from sleep paralysis and REM sleep behavior disorder to narcolepsy,
nightmares, and post-traumatic stress disorder (PTSD). In this chapter, we bring all of these disorders together, expanding on what we've
said before, showing how they are similar and how they differ, and
discovering what they can tell us about NEXTUP.

POST-TRAUMATIC STRESS DISORDER NIGHTMARES

*He heard the chopper before he saw it. But he knew it was too late. Larry
and Carlos, ten yards to his left and right, were already dead, and he
should be, too. As the chopper roared over him and began dropping
explosives on the enemy positions, he saw an arm lying halfway between
him and Larry. "It must be Larry's," he thought, "or is it mine?" He
seemed to feel someone shaking it, shaking his arm. He opened his eyes
and saw his wife, gently shaking him. "It's just a dream," she told him.
For what, the thousandth time?*

The dreams of people suffering from PTSD are not unique in the intensity of the terror that suffuses them, but rather in their recurrence and realistic replay of the actual traumatic event. We discussed PTSD in Chapter 5, where we suggested that PTSD is a disorder of sleep-dependent memory evolution: It reflects specific failures of sleep both to weaken emotional responses to traumatic memories and to promote their integration with older memories. At the time, we made only passing reference to the unique form of PTSD dreams. But much as the nature of PTSD during wakefulness gave us insights into the mechanisms and functions of memory evolution in that chapter, so too do PTSD dreams offer insight into NEXTUP and how it works.

One feature of dream content that we discussed in previous chapters is their associative and often metaphorical nature. Our dreams do not normally replay memories from our waking life. Instead, we dream about what happened when we were awake.

Magdalena Fosse, a Norwegian graduate student in Bob's laboratory, tested this idea that we dream *about* what actually happened rather than dreaming an accurate replica of the memory.[1] For two weeks, participants in her study wrote down their dreams each morning as soon as they woke up. They then underlined any parts of the dreams that they felt they knew the waking source of, and then they wrote descriptions of these sources.

Magdalena was trying to determine whether people actually replayed episodic memories in their dreams. We mentioned episodic memories in Chapter 8, describing them as memories of actual events in our lives; memories that we can bring back to mind in detail, essentially allowing us to relive the experiences. If we replayed these memories in our dreams, Magdalena argued, the dream would take place in the same location. What's more, it would have the same characters and objects, actions, emotions, and theme as in its waking source, because all of these features are bound together in episodic memories. So, for each underlined dream element, Magdalena asked her participants to indicate how similar the dream element was to its waking source—how well it matched the location, the characters and objects, the actions and emotions, and the theme of the waking event.

Here's the short answer: Few matches were found. Out of over 350 dream elements with identified waking sources, only 60—about one in six—even occurred at the same location in the dream as in the waking event. In the end, only five of these dream reports—less than 2 percent—met the requirements for replay of an episodic memory. This is what we would expect from NEXTUP. Although many aspects of sleep-dependent memory evolution appear to benefit from reactivation of the memories being processed, the network exploration that occurs during dreaming avoids reactivating the original, source memory. As we discussed in Chapter 8, NEXTUP is designed to find and activate weakly related, older memories that might suggest new uses and interpretations of the source memory. This is especially true in REM sleep, when the replay of episodic memories from the hippocampus is blocked, and higher levels of the neurotransmitter acetylcholine—together with a shutdown of noradrenaline—bias associative networks toward weaker associations. We also found that in REM sleep, semantic priming showed preferential activation of weak associates.

Thus the realistic replay of an episodic memory in a dream would reflect a failure of NEXTUP, and the repeated replay of such a memory would indicate malfunction of NEXTUP, at least in relation to that particular memory. And it's exactly what we see in PTSD sufferers, where replays of the actual trauma memory can appear in their dreams, often every night.

Having such "replicative nightmares" in the aftermath of a traumatic experience is a predictor of who goes on to develop PTSD, and repeated portrayal of trauma-related memories in nightmares is associated with more severe and more chronic daytime PTSD symptoms. In other words, when NEXTUP malfunctions, the brain's ability to naturally and automatically process trauma memories during sleep, without awareness or intent, becomes compromised. At the same time, many forms of sleep-dependent memory evolution become similarly compromised, especially those dependent on REM sleep. Here are some of the functions affected:

+ Stripping of peripheral details from emotional memories
+ Softening of emotional responses during future recall of memories
+ Integration of new memories with older, related memories
+ Extraction of gist and discovery of the meaning of memories
+ Network exploration to understand possible interpretations and uses of memories

As we saw in Chapters 6 through 8, all of these processes occur naturally during sleep. Why would they break down in PTSD? One possibility can be found in the body's normal response to stress. Whenever our body is stressed, either physically or psychologically, it responds by releasing stress hormones—including cortisol and adrenaline from the adrenal glands, and noradrenaline in the brain. During wakefulness, noradrenaline helps the brain bring potential threats into razor-sharp focus, blocking it from being distracted by irrelevant thoughts and sensations. The normal shutdown of noradrenaline release in REM sleep allows NEXTUP to search for weakly related memories. But if the hyperarousal of PTSD prevents this shutdown, NEXTUP is thwarted in this search.

Tom Mellman reported in 1995 that while measures of overall noradrenaline release dropped by 75 percent during sleep for healthy participants, it actually *increased* by 25 percent for those with PTSD.[2] This failure to suppress noradrenaline release can lead to the cascade of events shown in Figure 13.1. High levels of noradrenaline (NA), likely due to maintained hypervigilance, prevent the development of a fully functional REM sleep state and thereby prevent the normal suppression of episodic memory replay. High noradrenaline levels

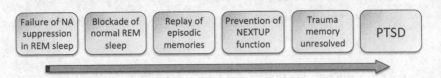

Figure 13.1. Steps in the development of PTSD following a trauma.

also limit the brain's ability to selectively seek out weaker associations. This constraint then blocks NEXTUP's ability to integrate the traumatic event into broader associative networks, and that integration is necessary to move past a traumatic experience. It's precisely this failure that defines the development of PTSD. Thus the breakdown of sleep-dependent memory evolution, including of NEXTUP, may be the ultimate reason that some trauma memories fail to evolve adaptively over time, and perhaps even reveal why PTSD develops.

But the neurobiology of dreaming and, by extension, of NEXTUP, is too complex to describe solely in terms of success versus failure. Between these extremities lies a vast middle ground where sleep-dependent memory evolution can take place in spurts and with varying degrees of effectiveness. For instance, although as many as 90 percent of trauma-exposed individuals who develop PTSD report nightmares bearing varying degrees of resemblance to the traumatic event, only about half of them experience these replays of the trauma memory in their nightmares. Instead, some post-trauma nightmares present distorted elements from the trauma, represent the traumatic event metaphorically, or replay the same distressing emotions experienced at the time of the trauma (such as terror, grief, helplessness) without direct references to the actual trauma event. Thus PTSD nightmares present a continuum of trauma-related replication tied to NEXTUP's degree of functioning. Over time, however, positive changes in the content of trauma-associated nightmares (such as decreases in the frequency and intensity of replicative elements, increased metaphorical representations of the traumatic event, and greater integration of everyday life events in dreams) tend to co-occur with clinical improvements in both overall mood and everyday functioning.

Scientists have yet to determine the extent to which progressive changes in dream content contribute to post-trauma recovery. But we believe that breakdowns in NEXTUP functioning—especially during the unique neurochemical and neurophysiological brain state of REM—are tied to important adaptive difficulties while its proper nightly functioning allows these difficult emotional memories to evolve effectively over time.

Prazosin

If development of PTSD begins with the increased levels of noradrenaline in the brain during sleep, what happens if you block the action of noradrenaline? Murray Raskind, a professor of psychiatry at the University of Washington in Seattle, stumbled on the answer to this question in 2000, when he was medical advisor to the Veterans Administration's Puget Sound African American Veterans Stress Disorders Program. While Raskind was there, two of his patients, both of them combat veterans of the Vietnam War who had PTSD, spontaneously and unexpectedly told him they had seen dramatic reductions in the severity of their PTSD nightmares, along with a return of more normal dreaming, after being treated with prazosin for an unrelated medical condition. Prazosin is an α1-adrenergic antagonist—a drug that blocks one of the many types of brain receptors that bind noradrenaline and mediate its action. Raskind, who had been studying noradrenergic abnormalities in patients with Alzheimer's disease, knew that some drugs activated these α1-adrenergic receptors and produced severe disruptions of sleep. He was also aware of Tom Mellman's findings of increased noradrenaline during sleep in patients with PTSD, as well as his report that these increased levels correlated with *decreases* in how long patients slept.

Maybe, Raskind thought, by reducing noradrenaline activity in the brain, prazosin was restoring the normal suppression of episodic memory replay in dreams. Indeed, prazosin had already been shown to reverse the suppression of REM sleep by drugs thought to increase noradrenaline in the brain. Since then, Raskind[3] and others have conducted a series of studies demonstrating prazosin's efficacy in reducing—and in some cases eliminating—these PTSD dreams. By decreasing the levels of noradrenaline in the brains of PTSD patients, prazosin restores the neurochemical environment required by NEXTUP to accomplish its work. Moreover, the reduction in chronic, trauma-related nightmares with prazosin is associated with clinically significant improvements in both sleep quality and daytime functioning. In the end, Raskind's reasoning was spot-on, and prazosin has become the most frequently recommended pharmacological treatment of trauma-related nightmares in PTSD sufferers.

IDIOPATHIC NIGHTMARES

As we saw in Chapter 10, idiopathic nightmares (those with no obvious cause) are ubiquitous; most people experience at least a few every year. Clinically significant nightmares, which usually occur at least once a week, are reported by about 4 percent of the general adult population. These nightmares are more common in women than men and are usually accompanied by significant daytime worries about the dreams. They are associated with a range of conditions including insomnia, depression, poor psychosocial adjustment, and suicidal ideation. But nightmares also occur in relatively well-functioning individuals. Where do these disturbing dreams come from, and why are they so prevalent?

There are likely as many beliefs about the origin of nightmares as there are about the nature and function of dreams more generally. Early explanations of nightmares centered on the idea of visitations from demons, ghosts, or other evil spirits. Contemporary explanations have focused on stress, unresolved conflicts, early childhood adversity, genetics, and nightmare-prone personalities. To further complicate matters, some researchers have proposed that nightmares serve a biological function; others believe they reflect a breakdown in normal function; and still others view them as unrelated to any biological functions whatsoever.

These divergences may be explained, at least in part, by the heterogeneity of nightmares. For any one individual, idiopathic nightmares can be acute and intermittent, or chronic and recurrent, and they can have varying contents. They may have emerged during childhood, adolescence, or adulthood, and they may feature the same primary emotion (such as helplessness) every time or may show a range of emotions. They can cause significant daytime distress and result in sleep-avoidant behaviors sometimes leading to chronic insomnia, or they can be perceived as inconsequential with little impact on everyday functioning. Thus, some of the factors commonly evoked to explain nightmares probably have a role in only some nightmares, experienced by only some people, and only some of the time.

The late Ernest Hartmann, whose ideas on the function of dreams we discussed in Chapter 7, spent much of his life studying nightmare sufferers. He found that many people with frequent nightmares, including lifelong nightmare sufferers, reported no clear history of childhood trauma and presented with no consistent pattern of psychopathology. But he also found that lifelong nightmare sufferers often had a personality type that included being unusually open and trusting, emotionally sensitive and creative. They were also likely to report unusual experiences such as dreams within dreams and déjà vu. Based on these observations, Hartmann[4] proposed that frequent nightmare sufferers had "thin psychological boundaries." He later developed the Boundary Questionnaire to measure this personality dimension, and research has borne out some of his ideas. When compared to people with "thick" psychological boundaries—typically described as being solid, well defended, rigid, and thick skinned—people characterized by thin psychological boundaries tend to remember more dreams, experience more nightmares, and report more intense and bizarre dreams.

Another factor that appears to contribute to nightmare occurrence is stress, although studies of this relationship have been inconsistent in their findings. One reason for this inconsistency is that stressors can come in many forms. There are acute stressors (exposure to war or natural disasters), experimental stressors (showing participants troubling films or having them complete a difficult "intelligence" test before going to sleep), emotional stressors (loss of work, divorce, death of a loved one), social stressors (interpersonal conflicts, loneliness, concerns about friends or colleagues), and just plain old daily hassles (being overworked, sitting in traffic, frequently misplacing keys or other objects). All of these stressors can build up over weeks, months, or even years.

And, of course, not everyone reacts to a given stressor in the same way. It's one reason that some people experience nightmares under periods of stress while others, exposed to the same stressor, do not.

We also know that *biological* markers of the stress response, such as levels of cortisol (the body's main stress hormone), can significantly differ from our *impression* of how stressed we feel. Our neurobiological reactions to stress—as distinct from our subjective perception of that stress—most likely also play a role in determining when we have nightmares.

Finally, our innate sensitivity to stressors—both experiential and biological—is partially controlled by our genetics. Indeed, one large-scale study[5] of nightmares in over 3,500 pairs of twins found gene variants associated with both childhood and adult nightmares. Exactly how these genes interact with environmental factors to affect the occurrence of our vividly emotional nightmares remains unanswered.

It's clear, however, that nightmares are common, multifaceted, and the product of a complex interplay between a host of psychological and biological factors, most of which we are just beginning to understand.

Idiopathic Nightmares and NEXTUP

If the replay of traumatic events in PTSD nightmares indicates a breakdown in NEXTUP functioning, and positive changes in the content of these dreams indicate at least partial recovery of NEXT-UP's ability to integrate traumatic memories into broader associative networks, where do non-trauma-related nightmares fit in? Part of the answer lies in how these nightmares unfold.

Back in Chapter 10, we described a study in which Tony and his then doctoral student Geneviève Robert analyzed the content of hundreds of bad dreams and nightmares. As part of that study, they also looked at how and when everyday dreams evolved into bad dreams and nightmares. They found that negative events external to the dreamer (for example, "I looked up at the sky and saw a missile coming right down at us") were the most common cause of everyday dreams turning into bad dreams or nightmares, occurring in about three-fourths of bad dreams and nightmares. In contrast, thoughts (such as "I was hovering high above a lake but then realized that if I was floating in the air, it was because I was dead") and emotions ("My sister walked into the room and I suddenly became very afraid of her") accounted for only a quarter of these dreams.

Almost invariably, these dreams begin innocuously. The dreamer describes the setting, presence of other characters, and mundane activities such as walking or looking around. But the nightmare trigger is usually not far behind; about 60 percent of the time, it appears during the first third of the dream. But just over a third of bad dreams and nightmares make it further into the dream before turning unpleasant. When Tony and Geneviève looked at the endings of these dreams, they found that about 20 percent of nightmares and almost 40 percent of bad dreams turn around and have either a partially positive ending (the dreamer escaped a danger but a partner was injured) or an entirely positive outcome (the dreamer took control of the situation or was saved at the last second). These findings provide insights into how NEXTUP operates in everyday nightmares and in dreams more generally.

Almost all dream reports start off by describing a relatively neutral or uneventful scene. Only after the dream gains momentum, as NEXTUP moves further along in its exploration of related memories, does the unfolding narrative take on a distinctly negative tone.

As we noted earlier, the dream event that turns an otherwise ordinary dream into a bad dream or nightmare usually originates in the dream world, although it is sometimes found in our own dream thoughts and feelings. This distinction is important because when brains dream, they create both the virtual world we become immersed in and the dream self who perceives and reacts to this simulated world from a first-person perspective. In Chapter 9 we discussed how NEXTUP explores associations that we would never normally consider when awake and then observes how our minds react to the resulting dream scenarios. But as we noted, the brain also sees how our ongoing thoughts, feelings, and actions, produced in response to this scenario, affect the people and events in the ever-shifting dream world. We also pointed out that it's during this remarkable interplay between our dream self and the rest of our simulated dream world that NEXTUP does its most important work. Unfortunately, it's also where bad dreams and nightmares have an opportunity to unfold.

As mentioned earlier, some people view dysphoric dreams as failures in some dream functions such as problem solving or emotion regulation. But others believe the exact opposite—that these dreams reflect the successful execution of dream function. From the perspective of NEXTUP, bad dreams and nightmares can reflect successes, failures, or anything in between, depending on the frequency and content of these dreams.

As an example, we know that emotions in nightmares can become so overpowering that they cause the dreamer to wake up, often with high levels of waking distress. Such events may happen because NEXTUP encounters a dead end while exploring novel associations to a particular event or emotional memory. This failure could result from a lack of weak associations to some memory that is truly outside an individual's experiences and understanding, or it could be a consequence of abnormally high levels of noradrenaline blocking access to weak associations, thereby thwarting NEXTUP activity. But many bad dreams simply reflect NEXTUP exploring unusually distressful associations, using them to integrate key memories within broader networks.

Even when a nightmare reflects a failure of NEXTUP, it isn't necessarily the end of the story. Just as people come back and reconsider intensely emotional events while awake, the dreaming brain almost certainly does the same, noting the circumstances under which powerful emotions are experienced and any events responsible for bringing the dream to a sudden end. A failure of NEXTUP in one dream can tag these memories for later exploration and even resolution in subsequent dreams. Of course, since we don't wake and recall most of our dreams, much of this revisiting will likely never be remembered in the morning. But we know that NEXTUP can revisit a troublesome topic over and over, across days, weeks, or months of dreaming. It's only when bad dreams and nightmares become chronic or repetitive that they are likely to reflect outright failures of NEXTUP and become associated with poor sleep and impoverished well-being.

NIGHTMARES AND IMAGERY REHEARSAL THERAPY

Many people with chronic nightmares believe there is nothing much they can do about these disturbing dreams, except maybe to avoid sleep altogether. In fact, only a minority of people with clinically significant nightmares ever discuss them with a healthcare professional, and less than one-third believe nightmares are treatable. And yet, they are. Earlier, we saw how prazosin became a recommended pharmacological treatment of trauma-related nightmares in PTSD sufferers. A safe and cost-effective nonpharmacological treatment for nightmares is imagery rehearsal therapy (IRT).[6]

Whether people suffer from trauma-related nightmares, recurrent nightmares, or chronic idiopathic nightmares, best-practice guides for the treatment of nightmares consistently recommend IRT, a cognitive behavioral intervention that teaches patients to rewrite their nightmares and rehearse the altered versions.[7]

Although IRT is administered in various ways, its core principle remains the same: nightmare sufferers "rescript" their nightmares in whatever way feels right to them and then rehearse the new dream through visual imagery (or, in the case of young children, through drawings) for several minutes each day. By instructing clients to "change your nightmare in any way that feels right to you," the therapist encourages nightmare sufferers to explore their own preferences for how to alter their dreams, as opposed to suggesting that they have to change the nightmare to something positive or triumphal. With this freedom, some people opt to change a minor detail in the dream, such as the color of a wall; others focus on giving the dream a new ending, and still others develop an entirely new story.

A number of studies have shown that IRT can reduce trauma-related—and unrelated—nightmares in everyone from children and veterans to trauma victims and patients suffering from mental illness. Importantly, gains achieved with IRT are maintained over time. What's more, successful treatment can lead to marked improvements not only in people's sleep and dreams, but in their waking life.

Although it's now clear that IRT is effective for most nightmare sufferers, it's still not clear why it works. One idea is that by rescripting their nightmares, people open up to the possibility that they *can* do something about their nightmares.

Several years ago, Tony collaborated on a study that looked at how female survivors of sexual assault rescripted their chronic nightmares with IRT.[8] He and his colleagues found that a consistent feature of these women's rescripted narratives involved gaining mastery over dream content by changing specific features of the dream—behavioral (fighting back or defeating an aggressor or other threat), social (having other dream characters help out), or environmental (changing the dream's hostile environment to a nonthreatening setting). In doing so, they created new associations to the trauma that the dreaming brain could use as stepping-stones in its exploration of alternate associative paths. Rescripting their nightmares during wakefulness also appeared to change how these women reacted to the nightmare *as it unfolded*, pushing it toward a more adaptive path. Whatever the mechanisms underlying IRT's therapeutic effects, the outcome is fewer disturbing dreams.

SLEEP PARALYSIS

Sleep paralysis, described briefly in Chapter 4, is among the strangest sleep-related parasomnias. As a reminder, it's characterized by the muscle paralysis in REM sleep continuing after you wake up, often with REM-like hallucinations that occur while you're awake with eyes open. The hallucinations typically include seeing a person—or some creature—in your bedroom, or even just *feeling* an evil presence nearby. About a quarter of adults have experienced sleep paralysis, so it probably doesn't qualify as extremely unusual; for some adults it's associated with narcolepsy, and for others it can be a recurring feature of their sleep.

Not surprisingly, the presence of hallucinated intruders in one's bedroom is confusing and often terrifying. You feel like you're dreaming, but you know you're awake. For hundreds of years, cultures have struggled to come up with their own understandings of these events. Sometimes the hallucinated creatures were thought to be demons, such as the one perched on the chest of a sleeping woman in Henry Fuseli's famous 1781 painting *The Nightmare*. At other times, they were thought to be angels. Even today, different cultures variably ascribe sleep paralysis to ghosts, witches, evil spirits, or attacks by dead bodies or unbaptized babies.⁹

Even in Western cultures, people who don't know about sleep paralysis struggle to come to terms with it. Bob had a laboratory assistant from Ireland who had initially gotten interested in studying sleep because of her own terrifying experiences with sleep paralysis when she was in college and going through a period when she wasn't getting enough sleep. At one point she brought her concerns to the attention of her priest; he even came to her house with his bible and holy water and performed an exorcism, in an attempt to get the evil spirit to leave. And, in a sign of the times, he also opined that it might be safer if she had her boyfriend sleep in bed with her. "I think my old college roommates and landlord still think I lost my mind," she told Bob.

Based on the history of this parasomnia, it's perhaps understandable that today many Americans experiencing sleep paralysis believe they have been abducted by space aliens. Richard McNally and Susan Clancy at Harvard have suggested that reports of alien abductions simply reflect one way our contemporary culture interprets the experience of sleep paralysis.¹⁰ Consider their description of one woman's experience:

> *[She] was lying on her back when she woke up from a sound sleep. Her body was completely paralyzed, and she experienced the sensation of levitating above her bed. Her heart was pounding, her breathing was shallow, and she felt tense all over. She was terrified. She was able to open*

her eyes, and when she did so, she saw three beings standing at the foot of her bed in the glowing light.[11]

It sounds like a typical experience of sleep paralysis; in an earlier time, she might have concluded that the three beings were angels or demons. Indeed, her first thought was that they were angels. But she was subsequently convinced by a friend that they must have been aliens.

Interestingly, none of the ten "abductees" whom McNally and Clancy interviewed thought at the time that they were being abducted. They only came to this conclusion later, after talking with friends or watching movies or TV shows about alien abductions. In its recruiting process, the research group did not mention sleep paralysis—the newspaper advertisements stated that researchers at Harvard University were "seeking people who may have been contacted or abducted by space aliens to participate in a memory study." Even so, all of these participants reported symptoms of sleep paralysis, which suggests that the majority of reported claims of such abductions result from attempts to explain the hallucinations of sleep paralysis. Of course, since a quarter of adults—more than 50 million Americans—report having experienced sleep paralysis, those who interpreted this event as an alien abduction are in the minority. But regardless, it's probably safe to say that these hypnopompic (relating to the transition from sleep to awakening) hallucinations are of no use for memory processing or NEXTUP, thus making it clear that sleep paralysis embodies a dream-related dysfunction.

REM SLEEP BEHAVIOR DISORDER

In Chapter 6, we briefly discussed REM sleep behavior disorder (RBD). It can be thought of as the opposite of sleep paralysis. Unlike REM sleep paralysis that creeps into wakefulness, RBD occurs when paralysis fails to develop in REM sleep. As a result, people with RBD are able to physically act out their dreams—in some cases dramatically—often injuring themselves or their bed partner.

It's rather curious that RBD was not identified until 1986. In that year, Carlos Schenck and Mark Mahowald in Minneapolis, Minnesota, reported a new category of parasomnia, seen in five patients they had identified over a two-year period. Four of the patients—all men—described acting out aggressive dreams that resulted in injuries to themselves or their spouses. In a second paper the following year, Schenck and Mahowald reported another five cases and named this new parasomnia REM sleep behavior disorder. Referring to the ten patients from these two studies, Schenck and Mahowald reported that "punching, kicking, and leaping from bed during attempted dream enactment caused repeated injury in nine of [these] patients."[12] Presumably RBD, which affects about one adult in every two hundred and particularly men over age fifty, has been around for hundreds of years. How had this phenomenon gone unnoticed?

We don't know for sure, but several factors likely contributed. First and foremost, it was probably assumed that these attacks and injuries occurred when the individual was awake, despite their insistence that they weren't. Women brought to the emergency room by their husbands who had attacked them during dream enactment were assumed to have been beaten by an abusive husband, and no amount of denial of this explanation by these women changed the diagnosis. The second factor is that until sleep recordings became available in clinical settings, there was no way to verify that these events were truly occurring during sleep. Remember, RBD was described just thirty years after REM sleep itself was discovered, when very few clinical sleep laboratories existed. But another factor in being slow to identify RBD is that it's a sleep disorder and, until comparatively recently, sleep just wasn't on the radar for most clinicians. After all, the thinking went, nothing interesting or important happens while people are asleep.

Unfortunately, even though episodes of RBD can be managed with medication, this disorder has a darker side. From the start, it was recognized that patients with RBD often had other neurologic disorders; five of the first ten people diagnosed with RBD in the 1980s displayed other neurologic symptoms. By 2012, the pattern was clear: more than 80 percent of patients with RBD go on to develop Parkinson's disease or other forms of dementia, and there's an average delay of fourteen years between the diagnosis of RBD and that of Parkinson's or dementia. This observation is perhaps not surprising, given that both RBD and Parkinson's disease are movement disorders that presumably result from deteriorations within the same brain networks that control movement. But RBD remains a harbinger of life-altering neurodegenerative diseases,[13] and though the symptoms of RBD can be effectively managed, we still don't have neuroprotective therapies that can delay or avoid the development of neurodegenerative disorders like Parkinson's disease in these patients.

NARCOLEPSY

Like sleep paralysis, narcolepsy is a sleep disorder in which the paralysis of REM sleep moves into wakefulness. We briefly described narcolepsy in Chapter 1, noting how almost all patients with narcolepsy report being confused about whether a dream had actually occurred in real life, a confusion that arose at least once a week for most of these patients. Then, in Chapter 4, we discussed the attacks of atonia, also called cataplexy, that cause narcolepsy patients to collapse limply to the ground despite being fully awake. Now let's dig a little deeper.

Unlike all the other sleep disorders we have described in this chapter, the cause of narcolepsy is precisely known. It arises from the death of a large proportion of the 10,000 to 20,000 neurons in the brain that produce the neurotransmitter orexin (also known as hypocretin). These might sound like large numbers, but not when you compare them to the 100 million neurons in each eye (about 10,000 times as many as those involved in making orexin) or to the 80 billion neurons in the whole brain (about 8 million times the number used in making orexin). The orexin neurons die as a result of an attack by the body's own immune system. Almost all narcolepsy patients with cataplexy have a specific mutation in an immune regulatory gene; that mutation causes the body to produce an antibody that attacks and destroys orexin neurons. Orexin controls the stability of our wake-sleep cycle, acting to keep us awake when we're awake and asleep when we're asleep. As a result, people with narcolepsy not only fall asleep frequently during the day but also wake up many times every night. Although medications and lifestyle modifications can help patients manage their symptoms, we still don't know how to halt or reverse the loss of orexin neurons and thus have no cure or prevention for the disorder itself.

Moreover, scientists have a good understanding of how this orexin deficiency affects sleep at the molecular, cellular, and brain network levels, but they don't understand its impact on dreaming. The delusional belief, *hours after waking up*, that what you're remembering is from a real waking event—when in fact it's from a dream—cannot be explained in terms of any direct effect of the orexin deficit. More research is needed to figure out this dream-related disorder.

SLEEPWALKING

Tony is lucky that his right hand isn't a mangled mess. Back when he was a graduate student, he got out of bed, dragged a wooden chair from the corner of his bedroom, stood on its armrests, and was about to reach into the spinning ceiling fan overhead when his girlfriend woke up and cried, "Tony! What are you doing!" Tony told her that they had to do something about the lobster infestation, and that he was going to start with the damn crustaceans dangling off the fan. Needless to say, there were no lobsters. It turns out that Tony, along with 2 to 4 percent of the general adult population, is a sleepwalker.

Tony's lobster anecdote illustrates three key features of sleepwalking, formally known as somnambulism. First, sleepwalkers are capable of surprisingly complex behaviors during their episodes. Besides standing precariously on a chair's armrests, sleepwalkers have been known to cook, eat (including horrendous mixes like peanut butter with pickles), rearrange furniture, get dressed, play musical instruments, wander outdoors, climb ladders, drive cars, and even wield weapons such as loaded shotguns. If that's not enough, there are documented cases of suspected suicide and even what has been accurately labeled homicidal somnambulism.[14]

Second, most sleepwalking children never remember their usually benign episodes. In contrast, up to 80 percent of adult sleepwalkers do, at least occasionally, remember "mini-dreams" related to their sleepwalking activities.

And third, although the behaviors of a sleepwalker may appear bizarre to an outside observer, they usually aren't random events. The actions of somnambulists can be driven by a sense of urgency or underlying logic even though the person's judgment is impaired. This feature is most famously seen in Shakespeare's *Macbeth*, when a guilt-ridden Lady Macbeth has a well-known sleepwalking episode: Crying, "Out, damned spot!" she tries to wash imaginary bloodstains from her hands, all the while talking in her sleep about the crimes she and her husband have committed.

Are sleepwalkers actually asleep during these episodes? Most sleepwalking arises out of deep N3 sleep. But unlike patients with REM sleep behavior disorder, somnambulists often appear to be awake; they are sufficiently aware of their environment to negotiate doors and stairs (and sometimes refrigerators) with relative ease. They can interact with people and, in extreme cases, even show surprising levels of self-awareness. One agitated sleepwalker sat up in bed and told his wife, "I know I sometimes have episodes, but this isn't one of them! There *is* an intruder in the house, and we have to get out!" In contrast, at other times sleepwalkers misperceive their environment, fail to respond to external stimuli, show signs of mental confusion, and, when awakened, have no recall of what they had been doing.

EEG studies[15] of sleepwalking episodes reveal a brain that looks both awake and asleep. Similarly, a brain imaging study of a sleepwalking episode showed that activity in some brain regions was turned down, just like it normally is during sleep. Meanwhile, activity in other regions was turned up to levels normally seen only in emotionally driven behavior during wakefulness. Clearly, sleepwalkers are neither fully awake nor fully asleep during these episodes. Instead, they are in a state characterized by the coexistence of sleep and wakefulness across brain areas. What's more, one study of sleepwalkers that Tony headed up showed that this coexistence of deep sleep and wakefulness can be seen up to 20 seconds *before* the episode actually begins.[16]

Are sleepwalkers *dreaming* during these events? That's a trickier question. For one thing, sleepwalkers are typically aware of their immediate physical environment while sleepwalking, which is not what happens in normal dreaming or in REM sleep behavior disorder. For another, sleepwalkers usually have their eyes open during these episodes, allowing them to navigate their surroundings. By contrast, normal REM and nonREM dreaming occur in a virtual, off-line world with exceedingly limited awareness of the real world. For these reasons, some people consider sleepwalkers' dreamlike perceptions to be closer to waking hallucinations than to bona fide dreaming. We doubt that this view is correct. As with sleep paralysis, sleepwalkers see a combination of the real and imaginary. In Tony's case, he perceived his actual ceiling fan but hallucinated the lobsters. A belief in a lobster infestation is, however, from the world of dreams and likely would never accompany a waking hallucination.

Sleepwalking has also been used as a tool to provide insight into memory processing during sleep. In one elegant study by Isabelle Arnulf and her colleagues in Paris and Geneva,[17] a sleepwalker who had been trained on a version of Bob's finger-tapping task—one that involved moving both arms through a sequence of large movements—

was filmed while enacting a fragment of the sequence in N3 sleep. These movements clearly reflected the memory reactivation during sleep that is the basis of most sleep-dependent memory processing, but with the effective motor activity of a sleepwalker. Was NEXTUP also at work? Unfortunately, no dream report was collected after this reenactment, so we'll never know if it was accompanied by a dream exploring the sleepwalker's associations to the task.

Epic Dreaming

Before we leave the world of dream-related disorders, let's consider one more. Imagine that every time you woke up, you felt exhausted, not because you slept poorly but because your nights were filled with long, tedious dreams of incessant physical activity such as repetitive housework or endlessly slogging through snow or mud. If this describes your nightly dreaming and ensuing daytime fatigue, you may suffer from *epic dreaming*. This disorder, a largely understudied form of disordered dreaming, was first described in 1995 by Schenck and Mahowald[18] (the same duo who first described REM sleep behavior).

Not much is known about this pattern of excessive dreaming other than that it affects women more than men. Sleep lab assessments usually come up clinically normal; and though the seemingly relentless dreams are followed by feelings of fatigue or exhaustion upon awakening, emotions within epic dreams are usually described as neutral or entirely absent. Even when epic dreaming occurs alongside nightmares, it is the impression of dreaming all night long that pushes these people to seek help. Psychological, behavioral, and pharmacological treatments for epic dreaming have proven largely ineffective. Why do epic dreamers feel the way they do, what does it all have to say about the function or dysfunction of NEXTUP, and what role do these patterns of dream recall and content play in the clinical symptoms of fatigue? All of these questions remain unanswered.

✧ ✧ ✧ ✧

IN CONCLUDING THIS CHAPTER, we'll make a few final points. First, sleep is an amazingly complex process that requires a wide range of brain systems to function in carefully coordinated patterns of activity. It's not surprising, then, that this unity of function sometimes breaks down. Second, many dream-related disorders are often greeted with superstitious fears and confusion, in much the same way mental disorders are. But we want to be clear: These are sleep disorders, not mental disorders. They arise, in every case, from a breakdown in the normal unity of action of the sleeping brain. Third, most dream-related disorders can be treated either with psychotherapy or medications. And finally, these fascinating disorders provide clinicians and researchers alike with powerful insights into the nature and function of dreaming.

Chapter 14

CONSCIOUS MIND, SLEEPING BRAIN

The Art and Science of Lucid Dreaming

WHEN TONY WAS EIGHTEEN, HE HAD A DREAM THAT changed his life. He dreamt that he was wrongly convicted of a crime, incarcerated, stabbed in the hand and chest by an inmate, and later escaped by sprinting across the prison yard and leaping over a 15-foot barbed wire fence while being fired at by sharpshooters. When he landed on a field of snow on the other side of the fence, he began to think something was wrong: The prison yard behind him (which now resembled his college) was covered in grass and not snow. And, come to think of it, that miraculous 15-foot jump didn't make much sense either. He examined his stab wounds. His flesh appeared to have healed, and the scaring pain was gone—as were the sharpshooters. Only one explanation made sense: he was dreaming.

Knowing that his real body was asleep in his bedroom, Tony picked up some snow in his dream, studied its granular texture, and marveled at the sensation of coldness permeating his hand. He then threw a snowball at the first person he saw—a large man standing about 20 yards away—curious about how he would react. The man, who now looked like a leather-clad biker, yelled at him and threatened to punch him out. When the biker took several steps in his direction, Tony panicked, forgetting for a moment that he was dreaming. He later encountered a couple of intriguing dream characters, and one of them tried to convince him that the whole thing *wasn't* a dream. By the time he finally woke up, Tony was enthralled and more than a little perplexed. He had experienced other dreams in which he knew he was dreaming, but never one so compellingly detailed and convoluted.

A few weeks later, while browsing in a second-hand bookstore, Tony picked up a book titled *Creative Dreaming*. Upon reading this book, written in 1974 by author Patricia Garfield, he discovered that his recent experience—knowing that he was dreaming while in the dream—had a name: *lucid dreaming*. Tony spent a good part of the following year reading up on REM sleep and dreaming and, after much thought, came to a decision. He wasn't going to follow in his brother's footsteps and go to medical school as he had originally planned. Instead, he was going to become a dream researcher; and as things worked out, he also became a pretty good lucid dreamer.

Like many people before him, Tony had been captivated by the experience and concept of lucid dreaming. Judging by the sheer number of books, websites, online discussion forums, apps, and popular articles dedicated to this unique form of dreaming, public interest in the topic has probably never been greater. Unfortunately, myths and misconceptions about lucid dreaming persist, and products claiming to allow people to "control their dreams"—mostly based on little to no evidence—abound. In this chapter, we take a lucid look at this fascinating aspect of dreaming.

WHAT LUCID DREAMING IS— AND WHAT IT ISN'T

The notion of lucid dreaming is relatively straightforward, but it is often misunderstood. Many people, for instance, assume that lucid dreaming is an all-or-nothing phenomenon—either you're aware that you're dreaming or you're not. In reality, dream lucidity exists on a continuum.

At one extreme are "pre-lucid" dreams, in which people question the soundness of their experience, even to the point of asking themselves, "Am I dreaming?" but mistakenly conclude that they aren't. (Darn, so close!) Then there's a kind of short-lived, low-level lucidity people sometimes experience at the end of a nightmare; they suddenly realize "Oh! This is a dream!" and promptly wake themselves up. Further along the continuum are dreams in which people experience a vague sense that they are dreaming or become lucid only in fits and starts; they either forget they are dreaming for extended periods of time or become so excited by this realization that they wake up. At the other extreme on the continuum are lucid dreams in which people not only know they are dreaming, but evince key mental abilities as if they were fully awake. They recall events from their waking life, are capable of logical reasoning, remember things they wanted to try in their dreams, and can consciously manipulate their physical actions in the dream. People having this kind of lucid dream can remember details such as what day of the week it is and where they are sleeping; they can willfully engage in supernatural actions like flying or passing through walls; and they can alter the course of the dream by, for instance, making objects appear or disappear. All of these experiences, from the barest hint that you're dreaming to conscious control over multiple facets of your dream's content, count as lucid dreaming.

Many people also confuse lucid dreaming with the notion of dream control. They are not the same thing. Although the two experiences can and often do occur at the same time, you can be aware that you are dreaming but also be incapable of or unwilling to alter the course of the dream; conversely, you can intentionally change things in your dream in ways that you couldn't in waking life without ever realizing that you are dreaming.[1] Moreover, the idea of dream control is something of a misnomer. Lucid dreamers can consciously direct their behaviors in their dreams, but most of them, at best, can only *influence* how the dream unfolds. For example, you may be able to make a character appear out of thin air in your dream, but good luck controlling what that person will say and do after materializing. The same goes for settings. You may decide to transport yourself to a nice outdoor café in Paris, but the scene will contain innumerable details that you didn't consciously choose or even consider, like whether the sky is clear or overcast, what the traffic is like, what the people around you are doing, and, for that matter, what clothes you are wearing.

Dreams, including lucid ones, are chock-full of details we don't normally pay attention to, let alone intentionally create or control. What's more, only about a third of lucid dreamers report being consistently able to manipulate their dreams, and even proficient lucid dreamers can have difficulty carrying out specific tasks in their dreams.

Unfortunately, because so much of our knowledge of lucid dreaming comes from self-reports, some people question whether lucid dreaming actually exists at all. Aside from subjective reports detailing such experiences, questions arise as to what objective, scientific evidence suggests that these dreams really do occur. It's a legitimate question. To many philosophers and scientists, the mere idea of self-awareness and reasoning during sleep seems to be an oxymoron. Sleep, after all, is characterized by the *loss* of conscious awareness of one's actual surroundings. It's no surprise, then, that many people, including some in the sleep and scientific community, have viewed lucid dreaming with skepticism, if not outright incredulity. But as we saw in our discussions of parasomnias such as narcolepsy, sleep paralysis, sleepwalking, and REM sleep behavior disorder, brain processes normally associated only with sleep or with wakefulness coexist in a number of fascinating conditions that have been scientifically studied. Lucid dreaming is just another one of these conditions.

LEFT-RIGHT-LEFT-RIGHT

In Chapter 2, we described the scanning hypothesis, which proposes that the rapid eye movements of REM sleep are related to gaze direction during dreaming. This hypothesis remains controversial, but it's clear that the eye movements recorded during REM sleep can correspond to where people subsequently report they were looking during specific activities in their dream, such as when climbing a ladder (upward eye movements) or watching a game of tennis (repetitive side-to-side eye movements).

In the mid- to late 1970s, these correspondences led Keith Hearne[2] at the Universities of Hull and Liverpool in England and Steven LaBerge[3] at Stanford University (both graduate students at the time) to independently come up with the same intriguing idea. Maybe lucid dreamers could mark the exact time they became lucid in a dream by making a series of agreed-upon eye movements (such as repeatedly looking far to the left and then far to the right) that would stand out on laboratory recordings of the electrooculogram, or EOG. Their hunch turned out to be spot-on. Alan Worsley, a skilled lucid dreamer and participant in Hearne's initial investigations, is credited with being the first lucid dreamer to signal when he became lucid in a dream. The original paper tracing of his ocular signals (showing his eyes going left, right, left, right, during a lucid dream in April 1975), along with the simultaneous EEG and EMG recordings proving that he was indeed asleep, are on permanent display in London's Science Museum.

Since then, dozens of laboratory studies have shown that lucid dreamers can communicate with researchers during periods of lucidity while in REM sleep by using predetermined eye-movement signals similar to the voluntary left-right-left-right (LRLR) eye movements just described. This method provides incontrovertible evidence of lucid dreaming, and it also allows lucid dreamers to "time-stamp" the exact moment they start and finish specific tasks in their dreams. A lucid dreamer may thus make one LRLR signal upon becoming lucid, a second when starting a predetermined activity, and a third when the action is completed. In this way, researchers know exactly *when* the sleeping participant was carrying out the experimental task during that dream (between signals two and three in our example) and thus how long the task took in real time. But it also tells researchers where to look in their psychophysiological recordings for corresponding changes in heart and respiration, muscle activity, EEG, and other bodily and brain functions. Lucid dream activities investigated in this way have included singing, counting, estimating time intervals, holding one's breath, clenching one's hands, doing leg squats, and even having sex.[4]

Think about this. You have a sleep researcher monitoring a sleeping participant who, once in REM sleep, remembers that they are asleep in a lab and that they have an experiment to carry out while in the dream. Then, using predetermined eye movement signals to reach out from inside the dream, they tell the researcher that they are now lucid and about to start a predetermined activity. And yet this isn't science fiction or some form of quackery—it's science.

What have these studies taught us about dreaming? Taken as a whole, these experiments suggest that the physiological changes that occur during lucid dreaming activities are the same as those observed when similar activities are carried out during wakefulness. For example, when lucid dreamers hold their breath in a dream, their bodies typically show a central apnea, a temporary cessation of breathing directed by the brain. When participants start to exercise in a lucid dream, their heart rate goes up. When one female lucid dreamer engaged in sexual activity in her lucid dream, researchers observed corresponding changes in multiple lab measures, including her respiration rate, skin conduction, and vaginal EMG. And in a case study using functional magnetic resonance imaging, clenching the hands during lucid REM-sleep dreaming and during wakefulness activated the same regions of the sensorimotor cortex in the brain.[5] Even though most of these findings are based on single case studies or very small sample sizes, they remain absolutely fascinating.

In our previous discussions of REM sleep behavior disorder, in which the paralysis that normally accompanies REM sleep breaks down, we saw that people with this disorder often physically act out their dreams. The studies of lucid dreaming add to this clinical literature in showing that as far as our brain is concerned, dreaming of doing something may not be all that different from actually doing it during wakefulness, even though overt muscle activity is blocked.

But is there a brain signature of lucid dreaming that might explain how self-reflectiveness arises during sleep? The answer appears to be yes, but with some important caveats.

WHEN BRAINS DREAM LUCIDLY

We've long known that most lucid dreaming occurs during REM sleep, especially later in the sleep period, when the sleeping brain shows higher levels of cortical activation. Beyond this finding, however, EEG studies of lucid dreaming have yielded mixed and largely inconsistent results. But recent brain imaging investigations present a slightly clearer picture.

When brains dream lucidly, frontal regions that are associated with self-reflective awareness during waking, but that are normally turned off during REM sleep, become more active. In fact, the communication among these frontal regions during lucid dreaming is increased to levels similar to those seen in waking.[6] But does becoming lucid lead to these changes in brain activity, or vice versa? At least one study suggests that it's the changes in brain activity that cause lucidity to develop. The study, led by Benjamin Baird in Giulio Tononi's laboratory at the University of Wisconsin–Madison, found that, compared to a matched control group of non-lucid dreamers, individuals who reported having at least three or four lucid dreams a week showed increased activity in frontal brain regions even when they were awake, resting with their eyes open.[7] These results suggest that the frontal increases in brain activity associated with lucid dreaming are present in lucid dreamers whether awake or asleep, and that this increased brain activity is facilitating lucidity in REM sleep.

Some researchers have tried to directly stimulate these frontal brain regions using electrodes pasted just above the forehead (a relatively recent method of brain stimulation known as transcranial electrical stimulation). But they've obtained only marginal evidence of a shift in dream content toward lucidity. In one study,[8] researchers devised a three-point scale where a score of 0 meant no evidence of lucidity, 1 reflected possible indications of lucidity, and 2 meant clear evidence of lucidity. The study team found that brain stimulation did increase scores, but only minimally—not even bringing them up to an average score of 0.5—and it did so only for those participants who reported normally having lucid dreams.

A second study, which many online and media outlets reported as being highly successful, didn't fare much better.[9] Although headlines proclaimed these researchers had shown that lucid dreaming could be triggered by stimulating the brain at specific frequencies, the results only showed that brain stimulation marginally increased participants' self-reported scores on measures of dissociation (such as taking on a third-person perspective in the dream) and insight (realizing they were dreaming). And not even one LRLR signal-verified lucid dream was collected. As for the boosted levels of insight, they were still less than one-fifth the level reported by lucid dreamers for actual lucid dreams.

Using yet another approach, researchers have tried to facilitate dream lucidity by strengthening REM sleep. One such study used the drug galantamine, which increases levels of acetylcholine in the brain. Acetylcholine regulates REM sleep, and galantamine not only increases the amount of REM sleep you get but also increases REM dream recall, vividness, and complexity. In the study, 121 participants were awakened after 4½ hours of sleep, given either galantamine or a placebo, spent a half hour practicing a mental technique to help induce lucid dreaming, and then went back to sleep. On nights when they were given a placebo, 14 percent of the participants reported having a lucid dream; on nights when they were given galantamine, almost half (42 percent) reported a lucid dream.[10] Although the results are remarkable, consider also these details of the study: First, participants were highly motivated adults taking part in an eight-day lucid dreaming workshop, and the study began on day five of the training program. Second, galantamine was combined with at least 30 minutes of sleep interruption, during which participants practiced a lucid dream induction technique before going back to sleep. Finally, galantamine can have side effects, including insomnia and gastrointestinal symptoms. Thus, while successful in inducing lucidity, this protocol doesn't seem like a method that many people would be eager to try.

But there is no end to the creative studies that researchers are carrying out. In collaboration with an international group of researchers, Ken Paller and his colleagues at Northwestern University are exploring ways in which researchers can communicate with lucid dreamers who are fast asleep in the laboratory. One approach consists of flashing lights through the closed eyelids of lucid dreamers after they have used LRLR signals to indicate that they are lucid while in REM sleep. The sequences of flashing patterns are used like Morse code and, as we saw in Chapter 2, these kinds of external stimuli can be incorporated into ongoing dream narratives. Lucid dreamers can recognize these signals and the coded message they carry and then respond to them using preestablished patterns of LRLR eye movements, thereby allowing two-way communication between the researchers and the lucid dreamers. Early results look promising, and the possibilities raised by these groundbreaking protocols are truly mind-boggling.

This new wave of research studies illustrates scientists' growing interest in understanding the neural underpinnings of lucid dreams, as well as the brain-based processes that facilitate their occurrence.

But public interest in such findings is often centered on how they can help people become lucid dreamers or, for those who already have lucid dreams, how to increase their frequency. And there is no lack of methods and devices claiming to help people achieve this goal.

LUCID DREAM INDUCTION

Given how much time we spend dreaming and how many dreams people remember, lucid dreaming is rather rare. Just a little over half the general population reports ever having had a lucid dream, and only 20 to 25 percent report having a lucid dream even once a month. Highly proficient lucid dreamers, who experience such dreams several times per week and can successfully produce LRLR signals in the sleep laboratory, represent far less than 1 percent of the population.

Most frequent lucid dreamers report either having had these dreams for as long as they can remember, or else having trained themselves to have lucid dreams at relatively young ages. Either way, almost all skilled lucid dreamers have excellent dream recall. If you want to become a lucid dreamer, you must develop good dream recall. As with the instructions given in Chapter 11 for eliciting problem solving dreams, if you want to learn to be a lucid dreamer, it's important to learn to remember your dreams when you wake up, and to keep track of the dreams you do remember with a diary, voice recorder, or app. The other important ingredient is motivation. Like many other learnable skills, becoming good at lucid dreaming takes time. But certain techniques can make it easier, and a growing number of wearable devices are aimed at short-circuiting the learning process.

It's an understatement to say that there is considerable public interest in these technologies. Over the past decade, several companies proposing lucid dream–inducing devices have launched wildly successful crowd-funding campaigns, often doubling or tripling their fundraising goals within weeks.

Most of these devices use sensors to detect electrical brain activity associated with REM sleep and, when the person is asleep and likely dreaming, they deliver stimuli such as lights, sounds, or vibrotactile sensations to alert the sleeper that they are dreaming. Other devices, similar to the transcranial stimulation studies described earlier, deliver electrical currents. Even though many of these devices will set you back a few hundred dollars, it's important to know that there is little to no published data on the efficacy of most of these products.[11] What's more, many of these companies have either gone out of business or experienced multiyear delays in delivering their products thanks to the many challenges involved in creating a device that can reliably induce lucid dreams.

In contrast, about three dozen studies have been published regarding the efficacy of various cognitive exercises designed to induce lucid dreaming.[12] Readers interested in these lucid dream–inducing techniques will find dozens of books and online resources dedicated to the task; approaches range from autosuggestion and visualization to techniques aimed at maintaining self-awareness while falling asleep. For the reader interested in giving it a try, here is a procedure that can help you get started.

STEPS TO BECOMING A LUCID DREAMER

1. Several times a day, ask yourself, "Am I dreaming?" Don't answer automatically; give it some thought. Take the time to examine your surroundings, to think about how you got there and what was going on right before you asked yourself this question. Fostering this attitude will help you recognize incongruities in your dreams and notice the sorts of memory lapses that often characterize dreams.

2. Get into the habit of asking yourself, "Am I dreaming?" whenever something surprising or improbable happens to you or when you experience powerful emotions. These are

the situations that are most likely to make you realize that you are, in fact, dreaming.

3. Perform daily reality checks, actions meant to determine if you are awake or dreaming. These can take many forms, including trying to read, staring at yourself in a mirror, turning on a light in a dark room, or trying to push your fingers through your palm. In most dreams, you will experience difficulties reading, mirror reflections quickly become unstable, light switches fail to work properly, and your fingers may pass through your palm or give rise to other unusual sensations. When carried out in a dream, these reality checks can help you realize that you're dreaming.

4. If you have a recurrent dream, or recurrent themes and settings in your dreams, rehearse these while imagining yourself realizing that you're dreaming.

5. Use the power of autosuggestion and tell yourself before going to sleep that you will have a lucid dream tonight; repeat it several times.

6. If you wake up in the middle of the night or early in the morning and intend to go back to sleep (or before taking a nap), take a moment and tell yourself that you will have a lucid dream or visualize yourself being lucid in a dream; you may enter REM sleep shortly after falling back asleep and be able to go straight into a lucid dream.

7. Remember that if you are truly unsure if you are awake or in a dream, you're almost sure to be dreaming!

These techniques can help you become a lucid dreamer or have more frequent lucid dreams (for some people, just reading this chapter may be enough to trigger lucid dreaming), but becoming lucid in a dream is the easy part. *Staying* lucid is the tough part. Maintaining dream lucidity requires walking a tightrope, always in danger of falling one way—back into non-lucid dreaming—or the other, forward into wakefulness. And learning to use dream lucidity to influence what happens next in a dream gets even trickier. In the following sections, we examine these and other higher-level aspects of lucid dreaming.

TO EXPECT, SPIN, AND PERHAPS TO DREAM

We've discussed the continuum of dream lucidity. And more often than not, a person's degree of lucidity will fluctuate within a given dream, even to the point of disappearing and then reemerging. It's not hard to see why. In Chapter 9 we detailed how dream plots, especially those arising from REM sleep, are rarely continuous across an entire dream. Instead, and as expected with NEXTUP, dreams frequently contain shifts in location, perspective, or action, and our dream thoughts similarly shift and sway with the dream's unfolding. Learning to keep mentally focused (to avoid falling back into non-lucid dreaming) and to modulate your emotions (to avoid waking up) as the dreaming brain stitches together an ever-evolving sequence of memories and network explorations is one of the biggest challenges faced by a lucid dreamer.

Being lucid and thinking clearly in dreams aren't the same thing. Even when lucid, a dreamer's reasoning is often clearly deficient. Sometimes, despite being lucid, dreamers incorrectly remember where they last fell asleep, forget that they don't need to surface to breathe while underwater, or reach outlandish conclusions like believing that the wise grasshopper talking to them really is an ancient god.

Once people realize that they are in fact dreaming, they'll often want to influence what happens next. Some lucid dreamers are able to make objects, people, or even entire settings appear or disappear at will, but most people find that *willing* things into existence just doesn't work for them. What does work is *expecting* things to happen.

In Chapter 9 we described how the ways we respond to various situations in our dreams can influence how they unfold, and in Chapter

10 we looked at how our dream thoughts, feelings, and actions can turn everyday dreams into bad dreams or nightmares. The same idea applies to lucid dreaming. Here's how you can do it.

Suppose you want to make a fancy house, delicious dessert, or particular person appear in your lucid dream. Instead of trying to force them to materialize before your very eyes, take a moment to imagine the desired object or scene. Then, slowly turn around in the dream while *expecting* to see the desired object behind you. Chances are that the person or scene (or a fair resemblance) will be there. Likewise, if you find yourself inside an office building in your dream, calmly walk toward a room while telling yourself that the person you want to see will be waiting for you inside, or that they are standing right around the corner. You can also use this "expectancy effect" to make wanted objects appear in your dreams: just tell yourself that you'll find the objects inside a desk drawer, behind some furniture, or even inside your pants pocket. Some experienced lucid dreamers will open doors in their dreams while envisioning whatever they want waiting on the other side. Not surprisingly, this expectancy effect works only if you are convinced that you can do it. If you have any doubts, your dreaming brain will likely track your uncertainties and take the dream in unexpected and potentially unwanted directions.

Consider the following example from a friend of Tony's we'll call Sarah. When Sarah was a young girl, she'd often dream of a large, menacing wolf chasing her through a forest. When she told her father about the dream, he told her that dreams couldn't hurt people and that the next time she had the dream, she should face the wolf and tell it, "Stop! You can't hurt me. This is a dream!" A few weeks later, Sarah had the dream again but remembered what her dad had told her to say. As soon as she finished repeating her father's words, the wolf came in close, looked her in the eye, and growled "Is that so?" It then ripped off her arm with a swift, vicious snap of its jaws. Sarah woke up shrieking and was upset at her dad for having given her such terrible advice.

The problem with Sarah's dream, however, wasn't necessarily with the advice itself. In fact, many people report having learned to lucid dream as a way of conquering their childhood nightmares. The hitch here is that even if Sarah had said the right words in her dream, part of her probably had been scared of what the creature might do. (In retrospect, Sarah thinks this was indeed the case.) This underlying sense of fear was, in all likelihood, registered by Sarah's dreaming brain and contributed to the unnerving ending of her dream. A similar dynamic can be seen in people who, while in the midst of an enjoyable flying dream, start to question how it is that they are able to fly—and promptly tumble back to earth.

But just as negative dream thoughts and feelings can take our dreams in unpleasant directions, positive dream thoughts and emotions can help steer them into more enjoyable spheres. Experienced lucid dreamers use their sense of volition and intentionality to influence or explore their dream worlds in any number of creative ways.

Some even learn how to prolong these experiences when they sense that their dream is about to come to an end—for example, when visual aspects of the dream, such as color, clarity, and brightness start to fade rapidly. Not wanting to wake up yet, lucid dreamers have explored numerous ways of extending their lucid adventures. (Yes, people do study and report on such things.) Among the more popular methods for prolonging lucid dreams is spinning your dream body. You can simply stretch out your arms in the dream and spin like a top or a dancer. It doesn't matter how you spin; the important thing is to *feel* the sensation of spinning. Another popular technique is to rub your hands in your dream, an action many lucid dreamers find particularly effective in stabilizing dream imagery.

No one knows exactly why these methods work. Some researchers believe that by forcing your dreaming brain to create vivid hallucinatory corporeal sensations in your dream body, you keep it from tracking or switching to inputs from your real body (such as how your hands and limbs are positioned in bed). Tuning out your real body, in turn, helps keep you in the dream. But these are merely speculative ideas.

WHO ARE YOU, AND WHAT DO
I MOST NEED TO KNOW?

Among the more fascinating aspects of dreams in general, and lucid dreams in particular, are the people we encounter in our internally generated dream worlds. Some people appear as one-dimensional extras in our dreams, but others have such a sense of realness about them that they elicit all kinds of reactions from us. Through their facial expressions, tone of voice, choice of words, emotional gestures, and overall behaviors and demeanor, dream figures can pull us into arguments, convince us of the necessity of helping them with some strange plan, make us leave a room in disgust, or fall madly in love with them. They can make us feel angry, scared, confused, or deeply aroused. But dream characters can also behave as if they experience their *own* distinct thoughts and feelings. They, too, can appear to be genuinely happy, fearful, or sad. Sometimes they even seem to know things that we don't!

By intentionally asking dream figures specific questions and monitoring their reactions, lucid dreamers have the unique ability to explore how their dreaming brain instantiates dream characters. In one fascinating study of dream characters' apparent mental abilities, the late Paul Tholey, a German dream researcher and Gestalt psychologist, had nine proficient lucid dreamers ask people in their dreams to perform specific tasks such as writing, drawing, coming up with a rhyme, or solving math problems.[13] It turned out that the dream figures were willing to give these tasks a try, and some were surprisingly good at them. Except for math, that is.

When asked to solve even basic arithmetic problems, such as three times four, dream characters typically struggled with finding an answer, although the dreamers did not. A second study similarly found that about two-thirds of the dream characters' answers were wrong.[14] Even more interesting, however, were some of the unusual answers and reactions noted by the lucid dreamers. When asked to solve a math problem, one dream character started to cry. Another ran away. In a handful of cases, dream characters acted like the question was overly personal or that the answers involved were either too subjective or important to be shared.

To some, these admittedly unusual studies are nothing more than amusing oddities. But the results do highlight a remarkable, yet often overlooked, aspect of our nightly interactions with the characters we meet in our dreams. Even when lucid dreamers consciously *decide* to ask dream figures specific questions and *know* that they are dreaming and that the person before them is a creation of their imagination, they still have little to no idea of how the dream character will react to their questions. This sense of unpredictability is only magnified in non-lucid dreams. In fact, virtually all interactions that we engage in while dreaming unfold without our knowing what dream figures will say or do next. In other words, even when lucid dreamers' expectations are in play, dream characters often behave in unexpected ways, as if they were following their own ideas and intentions. And since the characters in your dreams are created by your dreaming brain, every time something like this happens to you in a dream, you are, in a very real sense, surprising yourself.

Some lucid dreamers, particularly novice ones, will tell dream characters that they aren't real just to see how they react. Most of the time, the dream figures will simply ignore the dreamer's statement, scoff at the suggestion, or become upset with the dreamer. For many experienced lucid dreamers, a more interesting approach to interacting with central dream figures is to ask them potentially insightful questions, such as Who are you? Who am I? How can you help me? What is the most important thing I should know (about you, myself, what awaits us)? Dream characters often reply with pure gibberish, but often they appear to take these questions seriously, and their answers can be surprisingly witty or insightful.

Once you've learned to become lucid in your dreams, you can try this for yourself and see what types of reactions your dream characters have in store for you. Keep in mind, however, that even in lucid dreams, you are never the driver behind the wheel; your dreaming brain is. Still, by interacting in novel and sometimes thought-provoking ways with dream figures and noting how your thoughts, feelings, and actions influence the unfolding of your dreams, you may well learn something about how your dreaming brain constructs your inner dream worlds—and in the process, discover something about yourself.

✧ ✧ ✧ ✧

IN CONCLUDING THIS CHAPTER, we'll make a few final points. First, lucid dreaming is frequently portrayed as something that almost anyone can easily learn. But most people unfamiliar with the experience develop this skill only after investing considerable time and effort.

Second, people often make lucid dreaming sound like it's the ability to make anything and everything happen with a snap of your fingers. For an overwhelming majority of lucid dreamers, however, influencing dreams in a sustained or stable manner is significantly more difficult than these statements suggest: up to half of lucid dreams show no evidence of any kind of dream control. What's more, even proficient lucid dreamers often prefer to "go with the flow" and explore dreams as they unfold naturally rather than actively trying to alter their course.

Third, although claims about the benefits of lucid dreaming abound (from helping you overcome phobias to promoting physical healing to solving complex, real-life problems), most of these claims are unsupported by either clinical or scientific evidence. We're not saying that such effects are impossible, but rather that they have yet to be properly studied or shown to be true. We do know that lucid dreaming can be a lot of fun, that it can be used to treat nightmares, foster creativity, and perhaps even improve motor skills. Not a bad start, but still far from the abundant virtues attributed to lucid dreaming.

Finally, and perhaps most important, lucid dreaming is not for everyone. There's usually no harm in having lucid dreams or wanting to learn how to have them, but some people—including lucid dreamers—may experience terrifying lucid dreams that they cannot control and that they struggle to awaken from.[15] Besides intense feelings of fear and helplessness, these dreams—also known as lucid nightmares—typically contain acts of violence directed at the dreamer. In many cases, these acts are committed by ghoulish or demonic creatures. References to these uniquely terrifying dreams have existed for centuries and were even detailed by Frederik van Eeden, the Dutch psychiatrist who, in 1913, coined the term *lucid dreaming.* Even so, we still know very little about who is likely to experience lucid dreams and why. In addition, although certainly much less worrisome, the experience of false awakenings, which often accompany lucid dreams, is deeply unsettling to some people.

Despite these drawbacks, lucid dreams are largely quite stimulating, exciting, and often eye-opening experiences. They have already yielded an array of fascinating laboratory-based results, and they are being investigated by a growing number of researchers using a range of innovative tools. Progress is also being made, though more slowly, in understanding and testing the possible clinical and everyday applications of lucid dreaming. Ultimately, lucid dreaming offers a unique window into self-exploration, often taking place through interactions with other dream figures. But don't expect much help with your math homework.

Chapter 15

TELEPATHIC AND PRECOGNITIVE DREAMS

OR *WHY YOU MAY ALREADY HAVE DREAMT OF THIS CHAPTER*

DREAMS OF MIND READING. DREAMS OF EVENTS unfolding far away. Dreams that foretell the future. It seems like everyone either has had one of these dreams or knows someone who has. Other than being of interest since time immemorial and well ingrained in our cultural ethos, telepathic and precognitive dreams are the experiences that raise the most questions—and passionate debates—about the stranger, seemingly inexplicable side of dreaming.

This is probably why a great deal of the emails we receive from people who have questions about dreaming involve paranormal dream phenomena. People particularly like to detail how their dreams have predicted future events, from plane crashes to bombings to natural disasters. Many of these people go out of their way to make clear how often they have these dreams and how they *know* that they will come true, and they then provide examples of things they dreamt that later happened. And, not infrequently, they will end their email with something along the lines of "How do you explain *that*?"

As you'll see in this chapter, numerous difficulties are involved in addressing these kinds of dreams scientifically, and although many paranormal dreams *can* be explained, others simply can't. If you've ever had one of these precognitive (seeing the future) dreams, or one involving telepathy (communicating with someone directly from mind to mind) or clairvoyance (observing events not actually perceivable), you know how hard it is to shake the feeling that something particularly mysterious—beyond the bounds of normal science—has happened. And you're not alone.

Back in the 1880s, investigators at England's Society for Psychical Research—an organization that exists to this day—were studying a variety of paranormal experiences. In 1886, three of the society's founders—including Frederic W. H. Myers, who in 1882 coined the term *telepathy*—published *Phantasms of the Living*, a pioneering work that detailed the authors' critical investigation of hundreds of cases of telepathy and apparitions.[1] The book included 149 cases of spontaneous dream telepathy, the vast majority of which occurred between people who were either relatives or friends. Over half of these reports involved themes of death, and telepathic dreams about someone in danger or distress were the second most frequent category. After examining factors like report accuracy, corroborative evidence, explanations involving chance, "subliminal" memories, fraud, and so on, the authors argued for the existence of genuine evidence of telepathy, both in dreams and in waking life.

Some leading scientists of the time, such as the American psychologist William James, praised the book and its authors; others were critical of its theoretical and scientific conclusions. As we'll discuss, the gulf separating those who think paranormal experiences merit investigation and those who think such accounts are rubbish is probably greater today than it has ever been.

Alfred Adler, the third founder of the psychoanalytic movement (after Freud and Jung), was one of those people who viewed paranormal experiences as nonsense. In contrast, both Freud and Jung dabbled in these arcane phenomena and wrote quite extensively about them. Freud spells out his attitude, both ambivalent and coquettish, toward telepathy in his 1922 paper, "Dreams and Telepathy."[2] In beautiful though conflicting detail, the paper begins with this statement:

> *Anticipations will doubtless be aroused by the announcement of a paper with this title ["Dreams and Telepathy"]. I will therefore hasten to explain that there is no ground for any such anticipations. You will learn nothing from this [paper] about the riddle of telepathy; indeed; you will not even gather whether I do or do not believe in the existence of telepathy.*[3]

Jung, in contrast, showed no ambivalence at all, proclaiming in a 1933 letter that "the existence of telepathy in time and space is still denied only by positive ignoramuses."[4] He clearly believed that one determinant of dream content was telepathy:

> *The authenticity of this phenomenon can no longer be disputed today. It is, of course, very simple to deny its existence without examining the evidence, but that is an unscientific procedure which is unworthy of notice.*
>
> *I have found by experience that telepathy does in fact influence dreams, as has been asserted since ancient times. Certain people are particularly sensitive in this respect and often have telepathically influenced dreams.*[5]

By 1944, some ten years after Jung's letter and twenty after his own "Dreams and Telepathy" paper, Freud had tilted more toward accepting telepathic (but not precognitive) dreams as real. In a paper titled "The Occult Significance of Dreams,"[6] he concluded that based on available information, "one arrives at a provisional opinion that it may well be that telepathy really exists."[7] In the end, he notes that "I have often had an impression, in the course of experiments in my private circle, that strongly emotionally colored recollections can be successfully transferred [telepathically] without much difficulty [and] the possibility cannot be dismissed of their reaching someone during sleep and being received by him in a dream."[8] However, these accounts of telepathic phenomena—despite being endorsed by the likes of Freud and Jung—amounted to little more than anecdotes unsupported by actual scientific evidence. And yet . . .

In 1962, Montague Ullman, whose group approach to dreamwork we discussed in Chapter 12, founded one of the first sleep laboratories in New York City. More important still, the Dream Laboratory of the Maimonides Mental Health Center, as it was originally known, was the first and only sleep laboratory dedicated to the experimental study of dream telepathy. Ullman was joined in 1964 by Stanley Krippner, a Wisconsin native, who became director of the renamed Dream Laboratory. Together these researchers spent a decade studying dream telepathy. Krippner, a recognized leader in psychological research, is a past president of the humanistic psychology division of the American Psychological Association. During a ten-year period, Ullman and Krippner published a series of papers purporting to demonstrate dream telepathy.

In their first published study,[9] Ullman and Krippner designated one participant as a "sender" who concentrated on a picture while another participant (the "receiver") slept in the next room. The sender then tried to send a mental image of the picture to the receiver. In their first experiment, the pictures were simple line drawings, such as a circle or a bow and arrow drawn by the sender after the receiver went to sleep, or pictures from magazines. The sender would then wake the receiver when they were in REM sleep and ask for a dream report, posing further questions to elicit details. In 14 of 22 experimental sessions, Ullman and Krippner found what they considered significant similarities between the picture the sender had been concentrating on and the receiver's dream report. Encouraged by these results, and with the establishment of the dream laboratory at Maimonides, the two scientists began performing more rigorous experiments.

In their second experiment, Ullman and Krippner replaced their line drawings and magazine pictures with classic paintings, such as

The Sacrament of the Last Supper by Salvador Dalí. Again, the sender would focus attention on one of a dozen paintings, selected at random, and the receiver would try to dream about it.

Later, one or more judges would read the dream report, look at the entire set of pictures, and then pick the picture that they thought best matched the content of the dream report. If this "best match" picture was indeed the one the sender had been concentrating on, it would suggest that the sender's thoughts had somehow been transmitted to the dreamer in the next room and gotten incorporated into their dream. Of course, the match could be totally by chance, but based on the rates at which such agreements occurred, Ullman and Krippner concluded that the agreements were quite likely not just by chance. Instead, the two researchers concluded, they were actually observing dream telepathy. As you can imagine, when they published their results, there was a storm of protests and dismissals, some arguing that the dreamer somehow might have seen which picture the sender had chosen and others arguing that the statistical methods used were inappropriate.

It didn't help convince skeptics when Ullman and Krippner published the results of another study in 1973, titled "An Experiment in Dream Telepathy with 'The Grateful Dead.'"[10] This experiment relied on the approximately 12,000 fans attending a series of six Grateful Dead concerts. The concerts were held in the Capitol Theater in Port Chester, New York, and the fans acted as senders for two receivers, Malcolm Bessent and Felicia Parise. Bessent was wired for recording and sleeping in the Maimonides Dream Laboratory, about 40 miles away in Brooklyn, New York. Parise was asleep at home in Brooklyn. It was a stunning experimental design. Most of the senders "were in states of consciousness that had been dramatically altered by . . . the music, by the ingestion of psychedelic drugs before the concerts started, and by contact with other members of the audience."[11]

During each concert, one of six art prints was selected at random and projected for 15 minutes onto a large screen over the heads of the band while they played. Before the print was displayed, a series of slides explained to the fans:

> You are about to participate in an ESP experiment. In a few seconds you will see a picture. Try using your ESP to "send" this picture to Malcolm Bessent. He will try to dream about the picture. Try to "send" it to him. Malcolm Bessent is now at the Maimonides Dream Laboratory in Brooklyn.

Both Bessent and Parise were awakened multiple times across each of the six nights for dream reports. Both of them also gave associations to each of their reported dreams the next morning.

Afterward, transcripts of both dreams and associations for each of the nights were given to a pair of judges, along with copies of the six art prints. The judges were instructed to evaluate the "degree of correspondence between the dreams and the art prints" on a scale of 1 to 100 points. Then, for each night, the art print with the highest average correspondence was identified and compared to the art print shown to the concertgoers.

In evaluating Parise's dream reports, the judges picked the print actually shown to the audience on only one of the six nights. But in matching Bessent's dream reports to the art prints, the judges were correct on four of the six nights. Because there was one chance in six of matching up the right art print, the judges' correct matching of just one of six for Parise suggested that her dreams had not been affected by the picture shown each night. But the correct matching on four out of six nights for Bessent's dreams was extraordinary: the matches had only a 1 percent chance of occurring by coincidence. The fans, Krippner and Ullman concluded, had successfully transmitted telepathic impressions of the prints to Bessent on some of the nights. (Krippner recently complained about how much attention this particular study has attracted, pointing out that he and his colleagues have done much more rigorously scientific studies—and in all fairness, they have.)

The studies of dream extrasensory perception (ESP) conducted at the Maimonides Dream Laboratory are the best known in the field—and contain some of the most striking results. But dozens of other investigations of paranormal dreaming have been carried out, most of them conducted after the work by Krippner and Ullman. Several meta-analyses (a method used to systematically assess previous research to determine any overall statistical effect across studies) suggest that judges may be able to use dream reports to correctly identify target materials more often than would be expected by chance.[12] In other words, according to these researchers, dream ESP has a small but demonstrable effect that justifies continued study of this phenomenon.

The experimental investigation of paranormal dreams, however, is fraught with conceptual and methodological difficulties. One experiment that demonstrates the problems inherent in this area of research was performed in 2013 by Carlyle Smith, a well-respected scientist who has played a major role in the study of sleep-dependent memory processing.[13]

In this study, Smith invited the 65 students in his Trent University class on the psychology of dreams—all of whom had already recorded and turned in two dream reports as part of a class—to participate in an experiment that "targeted" an unidentified illness of a person unknown to them. Those who chose to participate were shown a photograph of a woman and told to "incubate a dream" about her medical problem (the concept of dream incubation is detailed in Chapter 11). Neither Smith nor the students knew the middle-aged woman or anything about her illness—breast cancer that had metastasized to one leg—until after the experiment was completed.

In the end, only a dozen students felt that they might have dreamed about the woman, and they submitted these dreams for content analysis. Only then were Smith and the students informed of the nature of her illness. But before any of the dreams were read, a scheme was devised to objectively score the reports for references to related topics; namely, any mention of the torso, limbs, breasts, cancer, or clinical settings. As Smith had predicted, the dreams collected *after* the students were shown the woman and asked to dream about her ailment were scored as more related to her cancer than the dreams collected before, and this result had only a 3 or 4 percent probability of being just by chance.

But there's a problem with the design. Can you see it? If we simply asked you to guess what medical problem middle-aged women are most fearful of, what would you guess? Very likely you would guess breast cancer. So the finding that these students had more dreams related to the target themes *after* seeing the woman's photo and learning that she had a medical problem than they did *before* knowing the nature of the experiment is not all that surprising: we don't know whether this increased rate of related dreams had anything at all to do with clairvoyance or was just the result of thinking, either consciously or unconsciously, about breast cancer after learning about the ill woman.

In response to such criticisms, Smith designed a second, better experiment. This time he added a control group that was shown a computer-generated composite, but believable, image of a nonexistent woman. Again, the experimental group was shown a photo of a real woman with multiple life problems, and they had more dreams related to her problems after seeing the photo than before. The control group shown the image of a nonexistent woman, however, had no more life-problem dreams after seeing the faux face than they had before. For the experimental group, the new experiment had the same problem as the first experiment —guessing what life problem the woman was trying to cope with.

But the study's limitation is subtler for the control group of the new experiment. This group, unfortunately, was run a year after the experimental group. As a result, it's possible that control participants had heard from last year's students about the study, and so would be expected to dream about the woman's life problems both before and after being shown the faux woman and having the experiment officially described to them. It would have been much better if both groups of students had been part of the experiment at the same time and then were randomly shown either the real woman or the faux woman. That would have eliminated our objection to the experiment design, although we might have come up with others.

What are we to make of all this? At the start of the chapter, we mentioned that we regularly receive emails detailing all kinds of paranormal dream experiences. Over the years, Tony has asked about a dozen people who have insisted that they frequently have precognitive dreams to email him reports of such dreams as soon as they occur (thus establishing a date) and then to let him know when the foretold events actually happen. Most haven't sent him any dream reports, and the few who have done so are yet to follow up with a single email confirming that the prediction came true. We also mentioned that after describing their paranormal dreams, people often seem to want to put *us* to the test by asking, "How do you explain *that*?" So, here are a few explanations for these kinds of dreams.

The first and probably most common explanation for dreams of "things that come true" has to do with probability and memory bias. We all experience several dreams a night, and billions of people are dreaming every night, so it's likely that on any given night, hundreds of people are dreaming of a plane crash, a volcanic eruption or tsunami, or the death of a famous person. But no one writes to a friend or researcher to say, "I dreamt of a huge industrial explosion two months ago and guess what? I'm still waiting for it to happen!" We only remember—and talk about—the dreams that appear to have come true. Most of us forget the ones that didn't.

Still, although any single precognitive dream could be simply a matter of coincidence, it often seems obvious that such coincidences are so improbable that ESP is the more likely explanation. We have the same thought when we're sitting in a group of twenty to twenty-five people and discover that two of us have the same birthday. How likely is that? Well, if there are twenty-three people, the odds are better than fifty-

fifty. If you do the math, it's actually more likely that two people in the group have the same birthday than that no two share a birthday. The point is, what's intuitively unlikely often isn't, and that probably explains many if not most proffered instances of dream ESP.

In reality, the likelihood of a correspondence between a dream we had and a subsequent waking-life event happening purely by chance is probably even higher than people imagine. One reason centers on all of those dreams that we don't remember. We know that our brains store memories of many dreams that we don't recall in the morning. You wake up with no dream recall, get in the shower, turn on the water, and suddenly remember that you dreamt about being in the shower. Or later in the day you see a cat run out in front of a car and say, "Oh! I dreamt about a cat last night," and the entire dream comes back to you. (Coincidentally or not, both of these examples have happened to Tony and Bob.) Most likely, in proportion to the dreams that we recall spontaneously, we have many more dreams stored in our memories; but we don't recall our stored dreams unless some event reminds us of them. And we don't know how long these dream memories stay ready to be recalled if some similar event, like a natural catastrophe or a parent's death, occurs. Thus the number of dreams that could show such similarities by chance is, again, likely much higher than we realize.

Yet another possibility is that the often fuzzy nature of our dream recall ("There was something about a cat . . ."), together with the hyper-associative nature of dreams, might lead us to think in retrospect that a dream was about a topic, like the death of a parent, when it really didn't have much in common with it at all ("I remember I was feeling really sad, as if I'd lost something . . ."). When Bob submitted his paper on the Tetris study (described in Chapter 7), one reviewer raised this issue. He pointed out that people might have seen geometric shapes in their dreams—something that has been commonly described in hypnagogic dream reports—and when they woke up, decided these shapes must have been Tetris images. (That conclusion turned out to be unlikely; when Bob looked at hypnagogic dream reports from participants who hadn't played Tetris, they didn't have geometric shapes.) It would not be surprising to find that we sometimes subtly and unknowingly alter our recall of a dream to match something that happened the next day. Insofar as this kind of alteration does happen, it would again increase the likelihood of a dream seeming to be meaningfully related to a future event when in fact it's just a coincidence.

Perhaps the most interesting explanation for paranormal dreams, however, is one similar to Freud's suggestion of the "dreamer making unconscious estimates and guesses." From this perspective, you really are reading someone else's mind or seeing the future, but not by any kind of ESP.

When Tony was a graduate student, a neighbor of his told him how, earlier that day, she'd been walking up the staircase outside her second-floor apartment when one of the steps snapped under her weight, nearly causing her to suffer a bad fall. But what she really wanted to share with this young dream researcher was that just a few nights earlier, she had had a dream in which she had fallen right through that very same staircase! When Tony went over to see the broken step in question, he noticed that some of the wooden steps showed signs of rot near their outer edges, particularly around the metal rivets connecting them to the handrail. The neighbor insisted that she had never noticed the rot, or she would have had the stairs repaired instead of risking serious injury. But her brain likely did notice the rotting wood, and from here, it's not hard to see how her dreaming brain ended up exploring a possibility in relation to this "unconscious" memory.

This process of unconscious estimates and guesses, however, can be much subtler. Consider the following example. A recently retired uncle tells his niece (we'll call her Susan) that he played a great round of golf with his friends, but that the effort caused him some pain in his shoulder—yet another sign of aging. That night, Susan dreams that her uncle died unexpectedly of a heart attack, and the next morning she's stricken when she learns that he indeed had suffered a fatal heart attack during the night. Susan has just experienced a dream of clairvoyance.

Or maybe not. Pain in the shoulder is a classic sign of angina—*referred pain* that you feel when your heart isn't getting enough oxygen. Susan may have learned this at some point, even though she might not have thought about it for years, so she didn't think of it when her uncle mentioned his round of golf, or even when she learned about his heart attack. But the information was still stored somewhere in her brain, and in keeping with NEXTUP, her brain did what it's supposed to. It explored associative networks to understand possibilities, and finding this rather unsettling memory about referred shoulder pain, crafted it into her dream. Her dreaming brain had indeed predicted the future, but it did so through the brain mechanisms that give us NEXTUP. And, as is true for almost all dreams, her brain didn't tell her how or why it constructed this dream.

Even if Susan's uncle had died three days after the dream, Susan would probably still have interpreted her dream as magically predicting her uncle's death. It might have made her belief even stronger, since now her dream seemed to see into the future instead of just seeing at a distance. In fact, she probably could have dreamed about her father dying and still felt that it was a "message" from the future about her uncle's impending heart attack, just scrambled a bit. Recall what we said in Chapter 8 about the "felt meaning" of dreams, and how we are neurochemically driven to believe that they're meaningful. Because her uncle's death apparently confirms the importance of the dream, it's all too easy for Susan to leap to precognitive explanations.

But sometimes none of this feels like an adequate explanation. We've talked about Bob's dog lab dream before. To remind you, back in 1980, Bob had this dream:

I was in the dog lab again, and we had just cut open the dog's chest. As I looked down, I suddenly realized that it wasn't a dog; it was [his five-year-old daughter] Jessie. I stood there dumbfounded, not understanding how we could have made such a mistake. Then, as I watched, the edges of the incision came back together and healed, and I noticed that there wasn't even a hint of a scar.

Jump forward thirty years to the birth of Bob's second son, Adam. Adam was born with tetralogy of Fallot, a congenital heart defect that is responsible for "blue babies." To correct this otherwise fatal heart defect, Adam underwent corrective surgery at the tender age of four months. The surgeon cut open his chest to perform the open heart surgery. Fortunately, it was a total success, and Adam, now sixteen, has lived a normal life ever since. He was left with hardly any scar at all.

It was more than a year after Adam's surgery when Bob made the connection to his long-ago dog lab dream. Spooky! How likely was it that just by coincidence, he would have such a highly memorable dream that would so perfectly predict Adam's surgery thirty years later? Given the rarity of such congenital heart defects, not very likely. Despite the odds, Bob feels confident about saying it was just that—a coincidence. But that's because he can't accept the alternative, which is some kind of magical precognition. (And Tony agrees with him.)

In the end, the real problem is that if dream telepathy or the like actually exists, it's rare, more symbolic than specific, and unreliable. No one has been found who can do it consistently over years; it doesn't work in studies with targets like the lottery number; and there are no foolproof experimental designs that repeatedly show significant effects beyond what would be expected by chance.

But arguably, an even larger problem is the wholesale rejection of the concept by the vast majority of the scientific community. Even among sleep researchers, few have read any of the published studies on dream telepathy, and those who have tend to shrug their shoulders and say, "There must be something wrong with the experiment," or "I still don't believe it."

This kind of reaction isn't limited to the sleep community. In the summer of 2018, *American Psychologist*—the flagship peer-reviewed journal of the American Psychological Association—published a paper reviewing the data for parapsychological phenomena (also known as psi phenomena). Its authors concluded that the "evidence provides cumulative support for the reality of psi, which cannot be readily explained away by the quality of the studies, fraud, selective reporting, experimental or analytical incompetence, or other frequent criticisms,"[14] A rebuttal to the review article was later published in the same journal. In that article, the authors stated outright that they did not bother examining the data for psi. The reason given: "The data have no existential value";[15] they're irrelevant. The entire rebuttal can be summed up as follows: psi phenomena are impossible, and so none of these claims can be true. Case closed.

We mention this example to illustrate the attitudes and beliefs scientists sometimes adopt about topics like dream telepathy. Although a handful of researchers are willing to apply scientific methods to investigate anomalous phenomena, others show skeptical interest from the sidelines, and still others—as we just saw—refuse to consider the possibility that such phenomena exist, even when empirical evidence may suggest otherwise.

Some people believe that, given the topics being investigated, research on paranormal phenomena should be held to a different scientific standard. This stance is not unreasonable. The astronomer Carl Sagan famously said that "extraordinary claims require extraordinary evidence," a perspective also offered by the French mathematician Pierre-Simon Laplace and the British philosopher David Hume in the eighteenth and nineteenth centuries.

But what qualifies as an extraordinary or impossible claim depends partly on your knowledge and beliefs. History is filled with examples of ideas and claims that were once believed to be extraordinary but turned out to be true, or at least widely accepted by the scientific community: ideas of planetary motion, Mendelian genetics, electricity, quantum mechanics, and the notion of dreams playing a role in memory processing. Moreover, many familiar concepts—including consciousness itself—still defy scientific explanation and (this may come as a surprise to many) no one really knows what gravity *is*.[16] If anything, the history of science tells us that dogmatic certitude—about what we know or what we *think* we know—is not always warranted.

In closing this chapter, we'll emphasize what we said earlier. When our brain dreams, it can sometimes predict the future or show what's happening at that very time somewhere far away. Sometimes it does this because, consciously or unconsciously, we have information that allows our brain to calculate and literally envision the possibility of these events. At other times, it happens by pure coincidence. This possibility is increased by both the vagueness of dreams and the brain's bias toward finding meaning in them. Together these factors make it even more likely that we'll find connections between a dream and an event that we did not learn about until later—even though the brain did not, in fact, use these connections in constructing the dream. Unfortunately, we can't confidently say that it can happen, as it so often feels, by mental telepathy or precognition.

Finally, you should know that we haven't included this chapter on paranormal dreams because we've had them ourselves (Bob's dog lab dream notwithstanding) or because we are convinced of their reality. We've done so because researchers are actively trying to investigate these phenomena experimentally, and their findings are still controversial. For the record, Bob's take on the matter is similar to Adler's (most likely bunk) while Tony's is closer to Freud's (unlikely, but who knows!). And that's just fine with both of us.

WHAT WE KNOW, WHAT WE DON'T KNOW

WHAT WE MIGHT NEVER KNOW, AND WHY IT ALL MATTERS

IN CONCLUDING, LET'S LOOK BACK ON SOME HIGH-lights from earlier chapters and raise some final questions. First, we'll consider the rapid growth in dream-related ideas and discoveries, many of them presented in this book, and examine how they fit in a broader time perspective. Then we'll review some core features of NEXTUP, our new theory of dream function that arose from the writing of this book. And finally, we'll address some loose ends that we have left unwoven, some questions only half answered, and some of the challenges and excitement that lie ahead.

THE PEAKS AND TROUGHS OF DREAM RESEARCH

When Freud wrote *The Interpretation of Dreams* in 1899, he devoted his opening chapter to an extensive review of the pre-twentieth-century scientific literature on dreams. In other words, he fit a hundred years of dream research into a single chapter. Today, an equally thorough review of even a small slice of contemporary dream research could fill an entire volume. Hundreds of articles have been written on dream content, the neurobiology of dreams, dream recall, and dream-related disorders. And if we were to consider dreams as a whole, that number would run into the thousands.

The contrast between Freud's work and the multitude of studies that followed reveals the explosion of new dream research. But this flood of productivity and discovery has come in waves.[1] The first of these waves was primarily clinical; it began about a decade after Freud's publication of *The Interpretation of Dreams* and lasted until the late 1930s. A second wave, one that could be described as a veritable tsunami, arose out of the discovery of REM sleep in 1953. Aserinsky and Kleitman's landmark paper in the prestigious journal *Science* detailed this unusual stage of sleep and its close association with dreaming. The authors raised enticing questions and opened the door to new approaches to dream research. Thanks to the discovery of REM sleep, many scientists began turning their attention to the laboratory study of dreams.

But the promising equation of "REM sleep equals dreaming" proved to be overly simplistic. Although studies garnered interesting insights on the psychophysiology of REM sleep dreams, the work largely failed to meet expectations. Moreover, the disappointing results shattered hopes that laboratory investigations into REM sleep and the bizarreness of REM sleep dreaming would help scientists understand psychiatric disorders such as schizophrenia and psychosis. In many respects, over the two decades of dream research following the discovery of REM sleep, scientists didn't get much closer to answering fundamental questions about the nature and function of dreams than the nineteenth-century sleep pioneers had (discussed in Chapter 2).

In frustration, some dream researchers abandoned the field. As a result, much of the funding available for laboratory-based dream research, particularly in the United States, dried up. Hobson and

McCarley's activation-synthesis hypothesis published in 1977 gave rise in the 1980s to the popular view—also common in the general scientific community—that dreams were likely meaningless reflections of the quasi-random firings of the sleeping brain. These developments made the scientific study of dreams feel even less compelling. By the time the two of us had made dream-related questions a focus of our own research efforts, dream science was facing an uphill battle.

But progress has continued. Bolstered by renewed interest in the cognitive and phenomenological aspects of dreaming, dream research underwent a renaissance. There was a sharp rise in the number of scientists interested in the nature of consciousness, along with growing evidence that sleep— and maybe dreaming itself—played a key role in learning and memory. Development of new approaches to the clinical and everyday use of dreams, increased public awareness of the suffering associated with PTSD nightmares, and burgeoning interest in lucid dreaming all helped to revitalize the field.

By the start of the twenty-first century, the synergistic effects of all these factors had given rise to a new wave of interest in and excitement about dreams. Today we are pleased to report that dreams and dreaming are widely accepted as legitimate objects of scientific study. What's more, the number of clinicians, philosophers, experimental psychologists, and neuroscientists working on the how and why of dreaming is at an all-time high. Like other dream researchers, we couldn't be happier.

In this book, we have compiled a wealth of recent insights and discoveries about the sleeping brain and the nature of dreams, weaving in ideas and findings from a vastly broader literature. Our goal has been to show you why the human brain needs to dream, in the process offering new answers to all four of our original questions: what dreams are, where they come from, what they mean, and what they are for. We hope our efforts have convinced you of just how remarkable the dreaming brain is and explained why so much can be gained by studying its nightly creations.

FURTHER THOUGHTS ON NEXTUP
AND THE FUNCTION OF DREAMING

When your brain constructs a dream, it creates an amazingly comprehensive virtual world within your mind. It generates illusory sensory experiences that are often indistinguishable from those generated by your sensory organs when you're awake. But as Tony pointed out to his nephew, when your brain dreams, you see things without using your eyes and hear things without using your ears. Your brain also generates illusory movements without your muscles actually contracting or your body moving. It generates illusory pain.[2] It can even generate illusory orgasms in quadriplegics who are physically cut off from such sensations when awake.[3]

And your brain generates illusory emotions. You feel emotions that are not being expressed in your body; you experience your muscles tensing, the hairs on your arms rising, your skin sweating, your stomach knotting, and you feel fear, although none of these physical aspects of fear are actually occurring. The existence of this illusory world raises a profound question. If dreams lack "real" perceptual inputs from the external world, and if experiences in dreams are largely disconnected from the bodily processes that many philosophers, psychologists, and cognitive neuroscientists believe are necessary for conscious awareness, how does the brain produce the all-too-real feelings we experience in our dreams?

As explained by NEXTUP, we know that when your brain dreams, it does not merely create a "series of thoughts, images, and emotions occurring during sleep," but also accomplishes something dramatically more complex and extraordinary. As your dreaming brain activates the neural maps that underlie both your sense of self and your understanding of the world, you experience and constantly interact with a rich, immersive, and multifaceted sensory dream world as it unfolds over time, and you do so from a very personal, first-person perspective. Because your brain typically populates this dream world with people, creatures, pets, and other objects you can interact with, you also experience more socially oriented feelings such as envy, sympathy, camaraderie, shame, arrogance, and pride. Not only that, but when your brain dreams, it also tricks you into believing that the other people inhabiting your dreams are also experiencing such feelings. Hence, you dream of a spouse acting in a fit of jealousy, an angry boss unhappy with your work, a group of friends overjoyed by a long-overdue reunion, or the wickedness of the intruder who has come to kill you.

Pause for a moment and think about this phenomenon. We're all so used to dreaming, and it feels so familiar that we lose track of how truly extraordinary it is that our brains construct these wondrous dream worlds within our mind's awareness. Each of us, with our own particular thoughts, feelings, perceptions, and actions, can partake every night of a world with literally endless possibilities.

By creating both you and your illusory environment, your brain not only observes how your mind reacts to the situations depicted in your dreams but also depicts how your responses influence the people and events in your dream. This continually changing, dynamic interplay between your dream self and the rest of your dream world provides the perfect environment for your brain to explore associations that you would never normally consider during waking. It's here, in this magical dream world, that NEXTUP teaches you about yourself and the world you inhabit, using this dream world to explore your past and better prepare you for an uncertain future.

And all these activities of the sleeping brain point to one of the major strengths of NEXTUP. Besides incorporating recent findings from the fields of neurobiology, sleep, learning, and memory, NEXTUP takes into account and seeks to explain key aspects of the *experience* of dreaming. As detailed in Chapters 8 through 10, several predictions derived from NEXTUP align perfectly with contemporary descriptions of the formal properties of dreams as well as their specific content. NEXTUP also proposes a way of conceptualizing how and why current concerns become embodied in our dreams and how the dreaming brain goes about exploring weakly related associations and possibilities in relation to these concerns. Moreover, as we saw throughout the later chapters of this book, NEXTUP can help us understand key characteristics of different kinds of dreams, from prophetic dreams to nightmares and lucid dreaming. The model also helps explain how dreams can facilitate creativity and why they can be a source of personal insight. NEXTUP is also the first model to propose that dreaming serves different—albeit interrelated—functions across the stages of sleep. And, finally, through its neurocognitive and neurobiological underpinnings, NEXTUP can be extended to any other mammals that may experience some form of dreaming. We've summarized the main features of NEXTUP in the appendix of this book.

WHAT'S NEXT?

When Bob's daughter Jessie (whose dream of her duck marionette began our book) was starting to think about college, she told her father that she wanted to be an engineer. "Why not a scientist?" he asked. "Because," she replied with an ironic smile, "at the end of a day's work, I want to have *fewer* unanswered questions, not more!"

As scientists, we have grown comfortable with this quandary. We never really answer any question to our total satisfaction. Whenever we seem to answer a question, we find that it just raises many more. NEXTUP is a perfect example. Elaborating this model allowed us to address several questions about the nature and function of dreams. But we then had to add multiple passages that came about from our exploration of the new questions that NEXTUP raises. All par for the course and, for us, a big part of the excitement of science.

What are some of the key challenges that lie ahead? Although we've made considerable progress in understanding how dreams are put together, we still don't really know how the brain picks the memories to use in constructing a given dream. We don't know if the face of the stranger in our dream comes from stored memories, for which we've just forgotten the context, or are constructed on the fly by putting together individual features from an assortment of memories. We're not sure what guides the behaviors, feelings, and personalities ascribed to the characters in our dreams. We don't know how the narrative structure of the dream is woven into a whole, or how emotions are brought into these narratives. And we have no idea how this entire process rises to consciousness in the form of dreams.

Some of these questions—those related to faces, narrative construction, and selection of memories—will eventually be answered by science, and we've discussed recent neural network and brain imaging techniques that will undoubtedly help answer them. But the timeline for finding answers related to the phenomenology of dreaming, the actual *conscious experiencing* of dreams, is considerably longer. These are questions that philosophers and then neuroscientists have been trying to tackle for millennia; and, frankly, we have no idea when or how they will be answered. We're not even certain that we'll ever find answers.

None of these tougher questions, however, are unique to dream research. Rather, they belong to the larger fields of cognition and consciousness research. Recall something that happened to you yesterday and think about it for a second. Got it? Well, as scientists, we don't have much of an idea of how your brain did that, how it searched yesterday's memories, selected just one of them, and found the associations that defined what it meant to you. And we know nothing about how you became conscious of any of this information. If you think of dreaming as a special case of consciousness—an altered state of consciousness—it's clear that dream researchers are trying to answer questions about this special case that haven't yet been answered for consciousness in general.

Bob raised the same complaint twenty years ago when people asked him to explain how sleep-dependent memory evolution worked, when researchers had little idea how memory evolution worked during wakefulness. At least now, and largely thanks to the findings of sleep researchers, many more in the general memory research community are addressing these questions about memory processing during wakefulness. Perhaps that will be the case for consciousness research, too. Maybe dream research will be the vanguard of a larger exploration of consciousness. If so, you are part of that vanguard.

But the future also holds other, more troubling questions involving modern society as a whole. First, what are the personal and social implications of emerging technologies that seek to alter your nightly dreams, either by amplifying their sensory intensity, modulating the emotional experience within them, or inducing specific kinds of dream experiences, such as lucid dreams? There's a tremendous appetite for such technologies, but we know next to nothing about their possible consequences. To what extent would your sleep be altered by such technologies, and how might these changes affect the core functions of sleep, including emotional and memory processing? Would there be a risk of some kind of dream addiction? Could artificially induced dreams become a world in which some people would take refuge, withdrawing from the harsher realities of their waking lives? And by manipulating dream content, might people inadvertently block critical functions of the dreaming brain?

Looking further ahead in time, what happens if scientists do become able to record our nightly dreams? Granted, until we have a way of recording people's thoughts and fantasies during wakefulness, the odds of doing so for dream experiences remain slim. But, as we saw in Chapter 7, researchers have already made some progress in using computational techniques to reconstruct the broad outlines of people's dreams. If and when such technologies become perfected and available, would you want to record your nightly dreams? What about those of your spouse or children? Who should have access to these dreams, and how would you control who does? In all likelihood, as is true of most sleep trackers and other wearable biosensors, such commercial developments would rest on proprietary (and closely guarded) algorithms. How willing would you be to trust others—a high-tech company, the cloud, or a corporate computing server—with something as personal as your dreams? How will you feel when the suggestions that pop up when you go on the internet come not from your previous searches, but from last night's dreams?

These issues might seem to be far in the future, but they are worth considering today. Given the current pace of biomedical and technological advances, we may come to regret not stopping to think about these questions now.

ON THE MYSTERY AND MAGIC OF DREAMS

For some people, the idea of scientific research to extend our understanding of dreams threatens the beauty and magic of these truly wondrous events. We believe this book will help put such worries to rest. If anything, the scientific ideas and findings we've presented throughout this book reveal the many ways in which dreams are psychologically and neurologically meaningful experiences, and how dreamers, artists, clinicians, and scientists stand to gain from paying attention to dreams. For us, everything that we learn about how dreams are formed and what functions they serve simply increases our awe of the dreaming brain. It doesn't take away the mysterious sense of wonder that our dreams create; it magnifies it. We hope this book has helped create and enhance this sense of wonder in you, as well.

And we hope that NEXTUP has given you a better understanding of the nature and function of dreaming. NEXTUP suggests that the function of dreaming is to explain the past and predict the future, to discover what's "next up" in our lives. This is the brain's task while we dream. But to achieve this goal, the dreaming brain attempts only to show us what has been and what might be. It does so in the same way as a great painter or composer, novelist, or playwright might—by *showing* us what we cannot yet fully explain. Arguably, that is the function of art; and we believe it is also the function of dreaming. And like good art, dreaming enriches our life while helping to guide us. Unlike our muscles, our brain and mind never rest; they function ceaselessly across day and night. It is perhaps a quintessential irony that, in the end, the mind never really sleeps. It dreams.

Appendix

NEXTUP:
A MODEL OF HOW AND
WHY WE DREAM

Note: Descriptions of the sleep stages mentioned in this appendix, including N1, N2, and REM sleep, can be found in Chapter 4.

I. Dreaming is a unique form of sleep-dependent memory evolution, one that extracts new knowledge from existing information through the discovery and strengthening of unexpected and often previously unexplored associations.

    A. To do so, dreams explore associations that the brain would never normally consider during waking; they search for and, when found, strengthen novel, creative, and insightful associations that the brain calculates to potentially be of future use.

    B. The reduction (in N2) or absence (in REM) of brain noradrenaline facilitates the search for weak associations.

    C. Dreams are not meant to solve ongoing concerns, but rather to explore them and their possible solutions in order to better understand what they mean for the dreamer.

D. Dreams normally show little transparent relevance to or usefulness for ongoing concerns. Rather, they identify unexpected associations that the brain calculates will be of use in resolving these and similar concerns in the future.

E. Reduced (in N2) or absent (in REM) levels of serotonin create states in which the brain is biased toward accepting dream associations as meaningful and useful.

II. Not all waking-life experiences and events are equally likely to be incorporated.

A. When brains dream, they tend to select ongoing concerns that have emotional salience.

B. The concerns selected contain unresolved questions whose answers, the brain calculates, would be of future use.

C. The concerns don't have to be big ones; they can be as simple as not knowing what a comment overheard earlier in the day meant or what time a bus will leave the next day.

D. Concerns can be identified and tagged for dream processing by NEXTUP during an actual event or when they surface during mind wandering or daydreams or even at sleep onset.

E. Dreams from sleep onset (N1), N2, and REM incorporate different concerns and associations.

1. Hypnagogic (N1) dreams tend to be overtly related to concerns that were being considered shortly before sleep onset.

2. N2 dreams tend to incorporate associations found in recent episodic memories, although less overtly.

3. REM dreams incorporate older and weaker semantic association with even less obvious relationship
to current concerns.

III. How the elements of dreams are brought together defines
their nature.

A. Dreams don't replay events from our life the way they're
replayed when we remember them during the day.
Instead, they tell stories about the event.

B. Dreams bring together fragments of both episodic and
semantic memories.

C. Complete episodic memories aren't incorporated into
dreams, and direct reference to or incorporation of current concerns is rare.

IV. The conscious experiencing of dreams is necessary to achieve
this goal.

A. The conscious experiencing of dreams is needed to create the narratives that allow the exploration of possible
scenarios.

B. It is also needed to produce the emotional feelings that
are critical for evaluating these scenarios.

C. It allows the brain to track how the dreamer's mind
reacts to the situations depicted in dreams, and, in
turn, to note how the dreamer's responses influence the
people and events in the dream.

V. Consequences of NEXTUP

A. Reduced serotonin levels during dreaming bias the
brain toward classifying weak associations not only as

useful but as meaningful; this probably explains why our dreams so often *feel* important.

B. Because the associations the brain incorporates into dreams are normally weak and previously unexplored, their connections to current concerns are usually not evident; even when such connections are identifiable, they are usually buried deep in a tangled narrative, obscured by the bizarreness commonly found in dreams.

C. Dreams do not need to be recalled after awakening to serve NEXTUP's function.

SUGGESTED READING

··

WE HOPE YOUR APPETITE FOR LEARNING MORE ABOUT
sleep and dreams has been whetted. If so, we have compiled a list of
resources that may be of interest to you. Our suggestions are few in
number and focus on what are, in our opinion, some of the best science-
based resources for the general reader. Some of these resources are also
listed among the dozens of books, scientific articles, and other references
cited throughout this book.

ONLINE RESOURCES ON SLEEP

National Sleep Foundation:
    https://www.sleepfoundation.org
Harvard Medical School Division of Sleep Medicine:
    http://healthysleep.med.harvard.edu/healthy
American Academy of Sleep Medicine:
    http://sleepeducation.org
Sleep on It! Canadian public health campaign on sleep:
    https://sleeponitcanada.ca

ONLINE RESOURCES ON DREAMS

International Association for the Study of Dreams:
  https://www.asdreams.org
Bill Domhoff and Adam Schneider's *The Quantitative Study of Dreams*:
  https://dreams.ucsc.edu
Kelly Bulkeley's *Dream Research & Education*:
  http://kellybulkeley.org

DREAM DATABASES

http://www.dreambank.net
http://sleepanddreamdatabase.org

GENERAL BOOKS ON SLEEP AND DREAMS

Given the many dozens of books on sleep and dreams that we have read and enjoyed over the years, and also the wide range of topics covered by these works, it is impossible for us to narrow our recommendations to a select few. Setting aside the many classics cited in this book, here are some of our recommendations.

Readers looking for insightful overviews of sleep and its importance are directed to Matthew Walker's *Why We Sleep* and William C. Dement's *The Promise of Sleep*. Readers interested in working with their own dreams should consider the works by Clara Hill (*Dream Work in Therapy*), Montague Ullman (*Appreciating Dreams*), and Gayle Delaney (*Living Your Dreams; Breakthrough Dreaming*). Those interested in the religious or spiritual dimensions of dreams can do no better than consult the works of Kelly Bulkeley (for example, *Big Dreams: The Science of Dreaming and the Origins of Religion*).

The number of books on lucid dreaming appears to be growing by the week, but we still think the original offerings by Stephen LaBerge

(*Lucid Dreaming*; *Exploring the World of Lucid Dreaming*) are hard to beat, although people interested in an in-depth, multidisciplinary exploration of lucid dreaming should consider the two-volume *Lucid Dreaming: New Perspectives on Consciousness in Sleep*, edited by Ryan Hurd and Kelly Bulkeley.

People interested in more academically oriented books on dreams are invited to examine any of the works by G. William Domhoff (*Finding Meaning in Dreams*; *The Emergence of Dreaming*) and the two-volume *Dreams: Understanding Biology, Psychology, and Culture*, edited by Robert Hoss, Katja Valli, and Robert Gongloff. Those interested in a more philosophical approach to dream research will be well served by reading Jennifer M. Windt's *Dreaming: A Conceptual Framework for Philosophy of Mind and Empirical Research*.

A Journey to the Far Side of Dreams

If you were intrigued by our discussion on the different ways people come to understand the concept of dreaming, or if you were captivated by the ideas presented in our chapter on lucid dreaming, you may be interested in reading Tony's recent suspense novel, *The Dreamkeepers*. Blending sleep science with dream mythology, this mystery thriller explores fictional dream worlds, delves into the forces that inhabit them, and takes the idea of lucid dreaming to new heights. Several people in Tony's dreams found it immensely entertaining. Maybe you will too.

ACKNOWLEDGMENTS

We are deeply grateful to many people, far more than we could possibly thank individually, who made this book and the novel ideas presented within it possible. Our heartfelt thanks go out to all of the research participants who helped further the science of dreams by keeping dream journals, completing questionnaires, sleeping in laboratories—sometimes under extraordinarily difficult conditions—or taking part in any number of other experiments in which their brain activity, thoughts, or emotions were studied in relation to their sleep and dreams. Likewise, much of the work conducted in our own respective labs wouldn't have been possible without the diligent and persistent efforts of dozens of graduate students, trainees, and technicians. In particular, Tony thanks Nicholas Pesant, Mathieu Pilon, Mylène Duval, Geneviève Robert, Aline Gauchat, Marie-Éve Desjardins, François White, Alexandra Duquette, Cristina Banu, Eugénie Samson-Daoust, Dominic Beaulieu-Prévost, Benoit Adam, and Dominique Petit. Bob thanks April Malia, Cindi Rittenhouse, Dara Manoach, David Roddenberry, Denise Clarke, Ed Pace-Schott, Erin Wamsley, Ina Djonlagic, Jason Rowley, Jess Payne, Magdalena Fosse, Margaret O'Connor, Matt Walker, and Sarah Mednick, his lab techs, grad students, postdocs, and colleagues.

Finally, our heartfelt thanks to our wonderful agent, Jessica Papin, and to Quynh Do at W. W. Norton and copyeditor Christianne Thillen for their excellent feedback and perspicacious editing.

From Tony: To Bob Pihl and Don Donderi I owe a special debt. Both were instrumental in helping me pursue my interests in dreams, first as a wide-eyed undergraduate student, then as part of my doctoral studies. I will always be grateful for their support, mentorship, and willingness to let me explore my (at the time) unusual research ideas. A great number of friends and colleagues played key roles in deepening my understanding of and appreciation for all things dreams. A special thanks goes to Bill Domhoff, Tore Nielsen, Rita Dwyer, Gayle Delaney, Alan Moffitt, Harry Hunt, Joseph De Koninck, Daniel Deslauriers, Anne Germain, Jacques Montplaisir, Carlyle Smith, Mark Blagrove, Jim Pagel, Ernest Hartmann, Ross Levin, Jacques Montangero, Isabelle Arnulf, Michael Schredl, Katja Valli, Mark Mahowald, Carlos Schenck, and Tracey Kahan. I also thank the members of the International Association for the Study of Dreams for their annual conference. Beginning in the late 1980s, these delightfully eclectic events allowed me to meet dozens of people whose passion for dreams was as remarkable as it was energizing. Thanks go to the Social Sciences and Humanities Research Council of Canada as well as the Canadian Institutes of Health Research for funding my research. A special thank you goes out to my parents, particularly my mom, who thought that studying dreams was a fine alternative to medical school. Thank you both for your encouragements and unwavering support. I also thank my two awesome boys and my amazing wife, Anne, for their incredible patience and support. Finally, my deepest thanks go to my longtime friend and now collaborator, Robert Stickgold. Bob, I knew that working with you would be intellectually stimulating, but I never thought it was going to be so much fun. Your broad-mindedness, scientific rigor, creative insights, and willingness to explore new ideas while revisiting old ones turned this exciting project into an extraordinary adventure into the heart of how and why we dream.

From Bob: My path to writing this book was paved by many people, starting with my sixth-grade teacher, Mr. Hampton, who turned me into a scientist, and my high school biology teacher, Fred Burdine, who turned me into a biochemist. Frank Neuhaus at Northwestern University was my first true mentor, who taught me how to be a scientist; Steve Kuffler at Harvard turned me into a neurobiologist; and Allan Hobson honed my skills as a dream researcher. I suspect that if any one of these people had been absent from my life, I would not be writing this now. Thanks go to Bob Rose at the McArthur Foundation and to the National Institute of Mental Health at the NIH, who provided both financial and moral support for my research. I owe a debt to many of the same dream researchers that Tony listed above, as well as Rosalind Cartwright, Ray Greenberg, and Milt Kramer. And I owe a huge thanks to my wife, Debbie, whose love and support sustained me throughout the writing of this book. Finally, I won't try to outdo Tony's thanks to me. So, I'll just offer an old childhood quote: Tony, "same to you, and double, too."

NOTES

CHAPTER 1: THINKING ABOUT DREAMS

1. Examples include Kelly Bulkeley, ed., *Dreams: A Reader on Religious, Cultural, and Psychological Dimensions of Dreaming* (New York: Palgrave, 2001); Robert L. Van de Castle, *Our Dreaming Mind* (New York: Ballantine Books, 1994), 3–106; and W. B. Webb, "Historical Perspectives: From Aristotle to Calvin Hall," in *Dreamtime and Dreamwork: Decoding the Language of the Night*, ed. Stanley Krippner (Los Angeles: J. P. Tarcher–St. Martin's Press, 1990), 175–84.

2. Monique Laurendeau and Adrien Pinard, *Causal Thinking in the Child* (New York: International Universities Press, 1962).

3. Laurendeau and Pinard, *Causal Thinking in the Child*, 106.

4. E. Wamsley, C. E. Donjacour, T. E. Scammell, G. J. Lammers, and R. Stickgold, "Delusional Confusion of Dreaming and Reality in Narcolepsy," *Sleep* 37 (2014): 419–22.

5. J. F. Pagel, M. Blagrove, R. Levin, B. States, R. Stickgold, and S. White, "Definitions of Dream: A Paradigm for Comparing Field Descriptive Specific Studies of Dream," *Dreaming* 11 (2001): 195–202.

CHAPTER 2: GRASPING AT DREAMS

1. J. Sully, "The Dream as Revelation," *Fortnightly Review* 53 (1893): 354–65.

2. Henri F. Ellenberger, *The Discovery of the Unconscious: The History and*

Evolution of Dynamic Psychiatry (New York: Basic Books, 1970); Frank J. Sulloway, *Freud, Biologist of the Mind: Beyond the Psychoanalytic Legend* (New York: Basic Books, 1983); P. Lavie and J. A. Hobson, "Origin of Dreams: Anticipation of Modern Theories in the Philosophy and Physiology of the Eighteenth and Nineteenth Centuries," *Psychological Bulletin* 100 (1986): 229–40; G. W. Pigman, "The Dark Forest of Authors: Freud and Nineteenth-Century Dream Theory," *Psychonanalysis and History* 4 (2002): 141–65.

3. Sigmund J. Freud, Jeffrey Moussaieff Masson, and Wilhelm Fliess, *The Complete Letters of Sigmund Freud to Wilhelm Fliess, 1887–1904* (Cambridge, MA: Belknap Press of Harvard University Press, 1985), 335.

4. Sigmund Freud, *The Interpretation of Dreams*, trans. J. Strachey (London: George Allen & Unwin, 1900/1954), 1.

5. Pigman, "The Dark Forest of Authors," 165.

6. Alfred L. F. Maury, *Le Sommeil et les Rêves: Études Psychologiques sur ces Phénomènes et les Divers États qui s'y Rattachent* [Sleep and dreams: Psychological studies on these phenomena and the various states associated with them] (Paris: Didier et Cie, 1861).

7. Karl A. Scherner, *Das Leben des Traums* [The life of dreams] (Berlin: H. Schindler, 1861).

8. Medard Boss, *The Analysis of Dreams* (New York: Philosophical Library, 1958), 25.

9. Freud, *The Interpretation of Dreams*, 359.

10. J. M. L. Hervey de Saint-Denys, *Les Rêves et les Moyens de les Diriger: Observations pratiques* [Dreams and the ways to guide them: Practical observations] (Paris: Amyot, 1867).

11. M. Calkins, "Statistics of Dreams," *American Journal of Psychology* 5 (1892): 311–43.

12. Calkins, "Statistics of Dreams," 312.

13. Sante de Sanctis, *I Sogni: Studi Psicologici e Clinici di un Alienista* [Dreams: Psychological and clinical studies of an alienist] (Turin: Bocca, 1899).

14. R. Foschi, G. P. Lombardo, and G. Morgese, "Sante De Sanctis (1862–1935), a Forerunner of the 20th Century Research on Sleep and Dreaming," *Sleep Medicine* 16 (2015): 197–201.

15. Sante de Sanctis, "L'interpretazione dei sogni [The interpretation of dreams]," *Rivista di Psicologia* 10 (1914): 358–75.

CHAPTER 3: FREUD DISCOVERED THE SECRET OF DREAMS

1. M. Kramer, "Sigmund Freud's *The Interpretation of Dreams*: The Initial Response (1899–1908)," *Dreaming* 4 (1994): 47–52.
2. Sigmund Freud, *The Interpretation of Dreams*, trans. J. Strachey (London: George Allen & Unwin, 1900/1954), xxv.
3. Freud, *The Interpretation of Dreams*, 233.
4. C. G. Jung, "Two Essays on Analytical Psychology," in *The Collected Works of C. G. Jung* (vol. 7), ed. Sir H. Read, M. Fordham, G. Adler, and W. McGuire (Princeton, NJ: Princeton University Press, 1967), 282.
5. Frederick C. Crews, *Freud: The Making of an Illusion* (New York: Metropolitan Books/Henry Holt, 2017).
6. J. F. Kihlstrom, "Freud Is a Dead Weight on Psychology," in *Hilgard's Introduction to Psychology*, ed. R. Atkinson, R. C. Atkinson, E. E. Smith, D. J. Bem, and S. Nolen-Hoeksema (New York: Harcourt Brace Jovanovich, 2009), 497.
7. Henri F. Ellenberger, *The Discovery of the Unconscious: The History and Evolution of Dynamic Psychiatry* (New York: Basic Books, 1970).
8. Sigmund Freud, "The Complete Letters of Sigmund Freud to Eduard Silberstein, 1871–1881," in *The Complete Letters of Sigmund Freud to Eduard Silberstein, 1871–1881*, ed. Walter Boehlich (Cambridge, MA: Harvard University Press, 1900), 149.
9. Sigmund Freud, Jeffrey Moussaieff Masson, and Wilhelm Fliess, *The Complete Letters of Sigmund Freud to Wilhelm Fliess, 1887–1904* (Cambridge, MA: Belknap Press of Harvard University Press, 1985), 417.
10. Freud, S., "Project for a Scientific Psychology," in *The Standard Edition of the Complete Psychological Works of Sigmund Freud, Volume I (1886–1899): Pre-Psycho-Analytic Publications and Unpublished Drafts*, ed. J. Strachey (London: Hogarth Press, 1895).
11. Benjamin Ehrlich and Santiago Ramón y Cajal, *The Dreams of Santiago Ramón y Cajal* (New York: Oxford University Press, 2017), 26.
12. Ehrlich and Cajal, *The Dreams of Santiago Ramón y Cajal*.

CHAPTER 4: THE BIRTH OF A NEW SCIENCE
OF DREAMING

1. E. Aserinsky and N. Kleitman, "Regularly Occurring Periods of Eye Motility, and Concomitant Phenomena, during Sleep," *Science* 118 (1953): 273–74.

2. W. Dement and N. Kleitman, "The Relation of Eye Movements during Sleep to Dream Activity: An Objective Method for the Study of Dreaming," *Journal of Experimental Psychology* 53 (1957): 339–46.

3. D. Millett, "Hans Berger: From Psychic Energy to the EEG," *Perspectives in Biology and Medicine* 44 (2001): 522–42.

4. M. F. van Driel, "Sleep-Related Erections throughout the Ages," *Journal of Sexual Medicine* 11 (2014): 1867–75.

5. A. Rechtschaffen and A. A. Kales, *Manual of Standardized Terminology, Techniques, and Scoring System for Sleep Stages of Human Participants* (Washington, DC: U.S. Government Printing Office, 1968).

6. H. P. Roffwarg, W. Dement, J. Muzio, and C. Fisher, "Dream Imagery: Relationship to Rapid Eye Movements of Sleep," *Archives of General Psychiatry* 7 (1962): 235–38.

7. H. S. Porte, "Slow Horizontal Eye Movement at Human Sleep Onset," *Journal of Sleep Research* 13 (2004): 239–49.

8. D. R. Goodenough, H. A. Witkin, D. Koulack, and H. Cohen, "The Effects of Stress Films on Dream Affect and on Respiration and Eye-Movement Activity during Rapid-Eye-Movement Sleep," *Psychophysiology* 12 (1975): 313–20.

9. T. A. Nielsen, "A Review of Mentation in REM and NREM Sleep: 'Covert' REM Sleep as a Possible Reconciliation of Two Opposing Models," *Behavioral and Brain Sciences* 23 (2000): 851–66; discussion 904–1121.

10. J. A. Hobson, *The Dreaming Brain* (New York: Basic Books, 1988).

CHAPTER 5: SLEEP—JUST A CURE FOR SLEEPINESS?

1. B. C. Tefft, "Prevalence of Motor Vehicle Crashes Involving Drowsy Drivers, United States, 2009–2013" (Washington, DC: AAA Foundation for Traffic Safety, 2014), https://aaafoundation.org/wp-content/uploads/2017/12/PrevalenceofMVCDrowsyDriversReport.pdf.

2. M. M. Mitler, M. A. Carskadon, C. A. Czeisler, W. C. Dement, D. F. Dinges, and R. C. Graeber, "Catastrophes, Sleep, and Public Policy: Consensus Report," *Sleep* 11 (1988): 100–109.

3. S. W. Lockley, L. K. Barger, N. T. Ayas, J. M. Rothschild, C. A. Czeisler, and C. P. Landrigan; Health Harvard Work Hours and Safety Group, "Effects of Health Care Provider Work Hours and Sleep Deprivation on Safety and Performance," *Joint Commission Journal on Quality and Patient Safety* 33 (2007): 7–18.

4. M. Lampl, J. D. Veldhuis, and M. L. Johnson, "Saltation and Stasis: A Model of Human Growth," *Science* 258 (1992): 801–803.

5. K. Spiegel, J. F. Sheridan, and E. Van Cauter, "Effect of Sleep Deprivation on Response to Immunization," *Journal of the American Medical Association* 288 (2002): 1471–72.

6. T. Lange, B. Perras, H. L. Fehm, and J. Born, "Sleep Enhances the Human Antibody Response to Hepatitis A Vaccination," *Psychosomatic Medicine* 65 (2003): 831–35.

7. K. Spiegel, R. Leproult, and E. Van Cauter, "Impact of Sleep Debt on Metabolic and Endocrine Function," *Lancet* 354 (1999): 1435–39.

8. L. Xie, H. Kang, Q. Xu, M. J. Chen, Y. Liao, M. Thiyagarajan, J. O'Donnell, D. J. Christensen, C. Nicholson, J. J. Iliff, T. Takano, R. Deane, and M. Nedergaard, "Sleep Drives Metabolite Clearance from the Adult Brain," *Science* 342 (2013): 373–77.

9. N. E. Fultz, G. Bonmassar, K. Setsompop, R. A. Stickgold, B. R. Rosen, J. R. Polimeni, and L. D. Lewis, "Coupled Electrophysiological, Hemodynamic, and Cerebrospinal Fluid Oscillations in Human Sleep," *Science* 366 (2019): 628–31.

10. R. Stickgold, J. A. Hobson, R. Fosse, and M. Fosse, "Sleep, Learning, and Dreams: Off-line Memory Reprocessing," *Science* 294 (2001): 1052–57.

11. M. P. Walker, T. Brakefield, A. Morgan, J. A. Hobson, and R. Stickgold, "Practice with Sleep Makes Perfect: Sleep-Dependent Motor Skill Learning," *Neuron* 35 (2002): 205–11.

12. J. D. Payne, D. L. Schacter, R. E. Propper, L. W. Huang, E. J. Wamsley, M. A. Tucker, M. P. Walker, and R. Stickgold, "The Role of Sleep in False Memory Formation," *Neurobiology of Learning and Memory* 92 (2009): 327–34.

13. D. L. Schacter and D. R. Addis, "Constructive Memory: The Ghosts of Past and Future," *Nature* 445 (2007): 27.

14. J. D. Payne, R. Stickgold, K. Swanberg, and E. A. Kensinger, "Sleep Preferentially Enhances Memory for Emotional Components of Scenes," *Psychological Science* 19 (2008): 781–88.

15. M.P. Walker and E. van der Helm, "Overnight Therapy? The Role of Sleep in Emotional Brain Processing," *Psychological Bulletin* 135 (2009): 731–48.

16. I. Djonlagic, A. Rosenfeld, D. Shohamy, C. Myers, M. Gluck, and R. Stickgold, "Sleep Enhances Category Learning," *Learning & Memory* 16 (2009): 751–55.

17. R. L. Gomez, R. R. Bootzin, and L. Nadel, "Naps Promote Abstraction in Language-Learning Infants," *Psychological Science* 17 (2006): 670–74.

18. D. J. Cai, S. A. Mednick, E. M. Harrison, J. C. Kanady, and S. C. Mednick, "REM, Not Incubation, Improves Creativity by Priming Associative Networks," *Proceedings of the National Academy of Sciences USA* 106 (2009): 10130–34.

19. R. Stickgold, L. Scott, C. Rittenhouse, and J. A. Hobson, "Sleep-Induced Changes in Associative Memory," *Journal of Cognitive Neuroscience* 11 (1999): 182–93.

CHAPTER 6: DO DOGS DREAM?

1. S. Coren, "Do Dogs Dream?" *Psychology Today*, October 28, 2010, https://www.psychologytoday.com/blog/canine-corner/201010/do-dogs-dream.

2. E. A. Lucas, E. W. Powell, and O. D. Murphree, "Baseline Sleep-Wake Patterns in the Pointer Dog," *Physiology and Behavior* 19 (1977): 285–91.

3. K. Louie and M. A. Wilson, "Temporally Structured Replay of Awake Hippocampal Ensemble Activity during Rapid Eye Movement Sleep," *Neuron* 29 (2001): 145–56.

4. E. Goode, "Rats May Dream, It Seems, of Their Days at the Mazes," *New York Times*, January 25, 2001, https://www.nytimes.com/2001/01/25/us/rats-may-dream-it-seems-of-their-days-at-the-mazes.html.

5. David John Chalmers, *The Character of Consciousness* (New York: Oxford University Press, 2010), 3.

6. T. Nagel, "What Is It Like to Be a Bat?" *Philosophical Review* 83 (1974): 435–50.

7. M. Grigg-Damberger, D. Gozal, C. L. Marcus, S. F. Quan, C. L. Rosen, R. D. Chervin, M. Wise, D. L. Picchietti, S. H. Sheldon, and C. Iber, "The

Visual Scoring of Sleep and Arousal in Infants and Children," *Journal of Clinical Sleep Medicine* 3 (2007): 201–40.

8. D. Foulkes, *Children's Dreaming and the Development of Consciousness* (Cambridge, MA: Harvard University Press, 1999); P. Sandor, S. Szakadat, and R. Bodizs, "Ontogeny of Dreaming: A Review of Empirical Studies," *Sleep Medicine Reviews* 18 (2014): 435–49.

9. Inge Strauch and Barbara Meier, *In Search of Dreams: Results of Experimental Dream Research* (Albany: State University of New York Press, 1996), 58–59.

10. Mark Solms, *The Neuropsychology of Dreams: A Clinico-Anatomical Study* (Mahwah, NJ: Erlbaum, 1997), 137–51.

11. E. Landsness, M. A. Bruno, Q. Noirhomme, B. Riedner, O. Gosseries, C. Schnakers, M. Massimini, S. Laureys, G. Tononi, and M. Boly, "Electrophysiological Correlates of Behavioural Changes in Vigilance in Vegetative State and Minimally Conscious State," *Brain* 134 (2011): 2222–32.

12. University of Liège, "Patients in a Minimally Conscious State Remain Capable of Dreaming during Their Sleep," *ScienceDaily*, August 30, 2011, https://www.sciencedaily.com/releases/2011/08/110815113536.htm.

13. B. Herlin, S. Leu-Semenescu, C. Chaumereuil, and I. Arnulf, "Evidence That Non-dreamers Do Dream: A REM Sleep Behaviour Disorder Model," *Journal of Sleep Research* 24 (2015): 602–609.

14. F. Siclari, B. Baird, L. Perogamvros, G. Bernardi, J. J. LaRocque, B. Riedner, M. Boly, B. R. Postle, and G. Tononi, "The Neural Correlates of Dreaming," *Nature Neuroscience* 20 (2017): 872–78.

15. "Animals Have Complex Dreams, MIT Researcher Proves," MIT News Office, January 24, 2001, https://news.mit.edu/2001/dreaming.

16. "Singing Silently during Sleep Helps Birds Learn Song," University of Chicago Medicine, October 27, 2000, https://www.uchospitals.edu/news/2000/20001027-dreamsong.html.

CHAPTER 7: WHY WE DREAM

1. T. Horikawa, M. Tamaki, Y. Miyawaki, and Y. Kamitani, "Neural Decoding of Visual Imagery during Sleep," *Science* 340 (2013): 639–42.

2. P. Maquet, J. Peters, J. Aerts, G. Delfiore, C. Degueldre, A. Luxen, and G. Franck, "Functional Neuroanatomy of Human Rapid-Eye-Movement Sleep and Dreaming," *Nature* 383 (1996): 163–66.

3. J. A. Hobson and R. W. McCarley, "The Brain as a Dream-State Generator: An Activation-Synthesis Hypothesis of the Dream Process," *American Journal of Psychiatry* 134 (1977): 1335–48.

4. R. W. McCarley and J. A. Hobson, "The Neurobiological Origins of Psychoanalytic Dream Theory," *American Journal of Psychiatry* 134 (1977): 1211–21.

5. Hobson and McCarley, "The Brain as a Dream-State Generator," 1347.

6. Hobson and McCarley, "The Brain as a Dream-State Generator," 1347.

7. F. Crick and G. Mitchison, "The Function of Dream Sleep," *Nature* 304 (1983): 111–14; 112.

8. A. Revonsuo, "The Reinterpretation of Dreams: An Evolutionary Hypothesis of the Function of Dreaming," *Behavioral and Brain Sciences* 23 (2000): 877–901; discussion 904–1121.

9. A. Zadra, S. Desjardins, and E. Marcotte, "Evolutionary Function of Dreams: A Test of the Threat Simulation Theory in Recurrent Dreams," *Consciousness and Cognition* 15 (2006): 450–63.

10. A. Revonsuo, J. Tuominen, and K. Valli, "The Avatars in the Machine—Dreaming as a Simulation of Social Reality," in *Open MIND*, ed. T. Metzinger and J. M. Windt (Cambridge, MA: MIT Press, 2016), 1295–1322.

11. Ernest Hartmann, *The Nature and Functions of Dreaming* (New York: Oxford University Press, 2010).

12. R. D. Cartwright, "Dreams and Adaptation to Divorce," in *Trauma and Dreams*, ed. Deirdre Barrett (Cambridge, MA: Harvard University Press, 1996), 79–185.

13. Owen Flanagan, *Dreaming Souls: Sleep, Dreams and the Evolution of the Conscious Mind* (New York: Oxford University Press, 2000).

14. David Foulkes, *Children's Dreaming and the Development of Consciousness* (Cambridge, MA: Harvard University Press, 1999).

15. G. William Domhoff, *The Emergence of Dreaming: Mind-Wandering, Embodied Simulation, and the Default Network* (New York: Oxford University Press, 2018).

16. R. Stickgold, A. Malia, D. Maguire, D. Roddenberry, and M. O'Connor, "Replaying the Game: Hypnagogic Images in Normals and Amnesics," *Science* 290 (2000): 350–53; 353.

17. Stickgold et al., "Replaying the Game," 353.

18. E. J. Wamsley, M. Tucker, J. D. Payne, J. A. Benavides, and R. Stick-

gold, "Dreaming of a Learning Task Is Associated with Enhanced Sleep-dependent Memory Consolidation," *Current Biology* 20 (2010): 850–55.

19. S. F. Schoch, M. J. Cordi, M. Schredl, and B. Rasch, "The Effect of Dream Report Collection and Dream Incorporation on Memory Consolidation during Sleep," *Journal of Sleep Research* (2018): e12754.

20. Wamsley et al., "Dreaming of a Learning Task."

21. G. W. Domhoff, "The Repetition of Dreams and Dream Elements: A Possible Clue to a Function of Dreams," in *The Functions of Dreaming*, ed. A. Moffett, M. Kramer, and R. Hoffmann (Albany: State University of New York Press, 1993), 293–320; 315.

22. A. R. Damasio, *The Feeling of What Happens* (New York: Harcourt Brace, 1999).

CHAPTER 8: NEXTUP

1. R. Stickgold, L. Scott, C. Rittenhouse, and J. A. Hobson, "Sleep-Induced Changes in Associative Memory," *Journal of Cognitive Neuroscience* 11 (1999): 182–93.

2. J. A. Hobson and R. W. McCarley, "The Brain as a Dream-State Generator: An Activation-Synthesis Hypothesis of the Dream Process," *American Journal of Psychiatry* 134 (1977): 1335–48; 1347.

3. G. W. Domhoff, "Dreams Are Embodied Simulations That Dramatize Conceptions and Concerns: The Continuity Hypothesis in Empirical, Theoretical, and Historical Context," *International Journal of Dream Research* 4 (2011): 50–62.

4. M. E. Raichle, A. M. MacLeod, A. Z. Snyder, W. J. Powers, D. A. Gusnard, and G. L. Shulman, "A Default Mode of Brain Function," *Proceedings of the National Academy of Sciences USA* 98 (2001): 676–82.

5. D. Stawarczyk, S. Majerus, M. Maj, M. Van der Linden, and A. D'Argembeau, "Mind-Wandering: Phenomenology and Function as Assessed with a Novel Experience Sampling Method," *Acta Psychologica* 136 (2011): 370–81.

6. M. F. Mason, M. I. Norton, J. D. Van Horn, D. M. Wegner, S. T. Grafton, and C. N. Macrae, "Wandering Minds: The Default Network and Stimulus-Independent Thought," *Science* 315 (2007): 393–95.

7. M. D. Gregory, Y. Agam, C. Selvadurai, A. Nagy, M. Vangel, M. Tucker, E. M. Robertson, R. Stickgold, and D. S. Manoach, "Resting State

Connectivity Immediately Following Learning Correlates with Subsequent Sleep-Dependent Enhancement of Motor Task Performance," *Neuroimage* 102, Pt 2 (2014): 666–73.

8. G. W. Domhoff and K. C. Fox, "Dreaming and the Default Network: A Review, Synthesis, and Counterintuitive Research Proposal," *Consciousness and Cognition* 33 (2015): 342–53; 345.

9. G. William Domhoff, *The Emergence of Dreaming: Mind-Wandering, Embodied Simulations, and the Default Network* (New York: Oxford University Press, 2018).

10. S. G. Horovitz, M. Fukunaga, J. A. de Zwart, P. van Gelderen, S. C. Fulton, T. J. Balkin, and J. H. Duyn, "Low Frequency BOLD Fluctuations during Resting Wakefulness and Light Sleep: A Simultaneous EEG-fMRI Study," *Human Brain Mapping* 29 (2008): 671–82.

11. C. J. Honey, E. L. Newman, and A. C. Schapiro, "Switching between Internal and External Modes: A Multiscale Learning Principle," *Network Neuroscience* 1 (2018): 339–56; 356.

12. Honey et al., "Switching between Internal and External Modes," 353.

13. E. J. Wamsley, K. Perry, I. Djonlagic, L. B. Reaven, and R. Stickgold, "Cognitive Replay of Visuomotor Learning at Sleep Onset: Temporal Dynamics and Relationship to Task Performance," *Sleep* 33 (2010): 59–68.

14. S. M. Fogel, L. B. Ray, V. Sergeeva, J. De Koninck, and A. M. Owen, "A Novel Approach to Dream Content Analysis Reveals Links between Learning-Related Dream Incorporation and Cognitive Abilities," *Frontiers in Psychology* 9 (2018): 1398.

15. A. S. Gupta, M. A. van der Meer, D. S. Touretzky, and A. D. Redish, "Hippocampal Replay Is Not a Simple Function of Experience," *Neuron* 65 (2010): 695–705.

CHAPTER 9: THE MISCHIEVOUS CONTENT OF DREAMS

1. Carolyn N. Winget and Milton Kramer, *Dimensions of Dreams* (Gainesville: University Presses of Florida, 1979).

2. Calvin S. Hall, *The Meaning of Dreams* (New York: Harper & Brothers, 1953).

3. Calvin S. Hall and Robert. L. Van de Castle, *The Content Analyses of Dreams* (New York: Meredith, 1966).

4. Calvin S. Hall and Robert. L. Van de Castle, "The Content Analyses of Dreams," dreamresearch.net, https://www2.ucsc.edu/dreams/Coding/.

5. A. Schneider and G. W. Domhoff, "DreamBank," www.dreambank.net.

6. C. Vandendorpe, N. Bournonnais, A. Hayward, G. Lachlèche, Y. G. Lepage, and A. Zadra, "Base de textes pour l'étude du rêve" [Text bank for the study of dreams], www.reves.ca.

7. D. Foulkes, "Dream Reports from Different Stages of Sleep," *Journal of Abnormal and Social Psychology* 65 (1962): 14–25; A. Rechtschaffen, P. Verdone, and J. Wheaton, "Reports of Mental Activity during Sleep," *Canadian Journal of Psychiatry* 8 (1963): 409–14; R. Fosse, R. Stickgold, and J. A. Hobson, "Brain-Mind States: Reciprocal Variation in Thoughts and Hallucinations," *Psychological Science* 12 (2001): 30–36.

8. K. Emmorey, S. M. Kosslyn, and U. Bellugi, "Visual Imagery and Visual-Spatial Language: Enhanced Imagery Abilities in Deaf and Hearing ASL Signers," *Cognition* 46 (1993): 139–81.

9. N. König, L. M. Heizmann, A. S. Göritz, and M. Schredl, "Colors in Dreams and the Introduction of Color TV in Germany: An Online Study," *International Journal of Dream Research* 10 (2017): 59–64.

10. J. Montangero, "Dreams Are Narrative Simulations of Autobiographical Episodes, Not Stories or Scripts: A Review," *Dreaming* 22 (2012): 157–72.

11. E. F. Pace-Schott, "Dreaming as a Story-Telling Instinct," *Frontiers in Psychology* 4 (2013): 159.

12. B. O. States, *Seeing in the Dark: Reflections on Dreams and Dreaming* (New Haven: Yale University Press, 1997).

13. M. Seligman and A. Yellen, "What Is a Dream?" *Behavioral Research and Therapy* 25 (1987): 1–24.

14. R. Stickgold, C. D. Rittenhouse, and J. A. Hobson, "Dream Splicing: A New Technique for Assessing Thematic Coherence in Subjective Reports of Mental Activity," *Consciousness and Cognition* 3 (1994): 114–28.

15. P. C. Cicogna and M. Bosinelli, "Consciousness during Dreams," *Consciousness and Cognition* 10 (2001): 26–41.

16. A. D. Wilson and S. Golonka, "Embodied Cognition Is Not What You Think It Is," *Frontiers in Psychology* 4 (2013): 58.

17. G. W. Domhoff and A. Schneider, "Much Ado about Very Little: The Small Effect Sizes When Home and Laboratory Collected Dreams Are Compared," *Dreaming* 9 (1999): 139–51; E. Dorus, W. Dorus, and A.

Rechtschaffen, "The Incidence of Novelty in Dreams," *Archives of General Psychiatry* 25 (1971): 364–68; C. Colace, "Dream Bizarreness Reconsidered," *Sleep & Hypnosis* 5 (2003): 105–28; Inge Strauch and Barbara Meier, *In Search of Dreams: Results of Experimental Dream Research* (Albany: State University of New York Press, 1996), 95–103.

18. E. J. Wamsley, Y. Hirota, M. A. Tucker, M. R. Smith, and J. S. Antrobus, "Circadian and Ultradian Influences on Dreaming: A Dual Rhythm Model," *Brain Research Bulletin* 71 (2007): 347–54.

19. C. D. Rittenhouse, R. Stickgold, and J. Hobson, "Constraint on the Transformation of Characters, Objects, and Settings in Dream Reports," *Consciousness and Cognition* 3 (1994): 100–113.

20. P. Sikka, K. Valli, T. Virta, and A. Revonsuo, "I Know How You Felt Last Night, or Do I? Self- and External Ratings of Emotions in REM Sleep Dreams," *Consciousness and Cognition* 25 (2014): 51–66.

21. T. A. Nielsen, D. Deslauriers, and G. W. Baylor, "Emotions in Dream and Waking Event Reports," *Dreaming* 1 (1991): 287–300.

22. M. Schredl and E. Doll, "Emotions in Diary Dreams," *Consciousness and Cognition* 7 (1998): 634–46.

23. Mélanie St-Onge, Monique Lortie-Lussier, Pierre Mercier, Jean Grenier, and Joseph De Koninck, "Emotions in the Diary and REM Dreams of Young and Late Adulthood Women and Their Relation to Life Satisfaction," *Dreaming* 15 (2005): 116–28.

CHAPTER 10: WHAT DO WE DREAM ABOUT? AND WHY?

1. C. Hall and R. Van de Castle, *The Content Analysis of Dreams* (New York: Appleton-Century-Crofts, 1966); D. Kahn, E. Pace-Schott, and J. A. Hobson, "Emotion and Cognition: Feeling and Character Identification in Dreaming," *Consciousness and Cognition* 11 (2002): 34–50.

2. G. William Domhoff, *Finding Meaning in Dreams: A Quantitative Approach* (New York: Plenum, 1996), 119–20.

3. R. M. Griffith, O. Miyagi, and A. Tago, "Universality of Typical Dreams: Japanese vs. Americans," *American Anthropologist* 60 (1958): 1173–79.

4. T. A. Nielsen, A. Zadra, V. Simard, S. Saucier, P. Stenstrom, C. Smith, and D. Kuiken, "The Typical Dreams of Canadian University Students," *Dreaming* 13 (2003): 211–35.

5. J. Mathes, M. Schredl, and A. S. Goritz, "Frequency of Typical Dream Themes in Most Recent Dreams: An Online Study," *Dreaming* 24 (2014): 57–66; F. Snyder, "The Phenomenology of Dreaming," in *The Psychodynamic Implications of the Physiological Studies on Dreams*, ed. H. Madow and L. Snow (Springfield, IL: Charles Thomas, 1970).

6. A. Zadra, "Recurrent Dreams: Their Relation to Life Events," in *Trauma and Dreams*, ed. Deirdre Barrett (Cambridge, MA: Harvard University Press, 1996), 241–47; A. Zadra, S. Desjardins, and E. Marcotte, "Evolutionary Function of Dreams: A Test of the Threat Simulation Theory in Recurrent Dreams," *Consciousness and Cognition* 12 (2006): 450–63; A. Gauchat, J. R. Seguin, E. McSween-Cadieux, and A. Zadra, "The Content of Recurrent Dreams in Young Adolescents," *Consciousness and Cognition* 37 (2015): 103–11.

7. G. Robert and A. Zadra, "Thematic and Content Analysis of Idiopathic Nightmares and Bad Dreams," *Sleep* 37 (2014): 409–17.

8. A. Zadra and J. Gervais, "Sexual Content of Men and Women's Dreams," *Sleep and Biological Rhythms* 9 (2011): 312.

9. M. Schredl, S. Desch, F. Röming, and A. Spachmann, "Erotic Dreams and Their Relationship to Waking-Life Sexuality," *Sexologies* 18 (2009): 38–43.

10. A. Zadra, "Sex Dreams: What Do Men and Women Dream About?" *Sleep* 30 (2007): A376.

11. D. B. King, T. L. DeCicco, and T. P. Humphreys, "Investigating Sexual Dream Imagery in Relation to Daytime Sexual Behaviours and Fantasies among Canadian University Students," *Canadian Journal of Human Sexuality* 18 (2009): 135–46.

12. M.-P. Vaillancourt-Morel, M.-È. Daspe, Y. Lussier, A. Zadra, and S. Bergeron, "Honey, Who Do You Dream Of? Erotic Dreams and Their Associations with Waking-Life Romantic Relationships," in *Great Debates and Innovations in Sex Research* (Montreal: Annual Meeting, Society for the Scientific Study of Sexuality, November 8–11, 2018), www.sexscience.org.

13. J. Clarke, T. L. DeCicco, and G. Navara, "An Investigation among Dreams with Sexual Imagery, Romantic Jealousy and Relationship Satisfaction," *International Journal of Dream Research* 3 (2010): 54–59.

14. J. B. Eichenlaub, E. van Rijn, M. G. Gaskell, P. A. Lewis, E. Maby, J. E. Malinowski, M. P. Walker, F. Boy, and M. Blagrove, "Incorporation of

Recent Waking-Life Experiences in Dreams Correlates with Frontal Theta Activity in REM Sleep," *Social Cognitive and Affective Neuroscience* 13 (2018): 637–47.

CHAPTER 11: DREAMS AND INNER CREATIVITY

1. Paul Strathern, *Mendeleyev's Dream: The Quest for the Elements* (New York: Hamish Hamilton, 2000), 286.

2. O. Theodore Benfey, "August Kekulé and the Birth of the Structural Theory of Organic Chemistry in 1858," *Journal of Chemical Education* 35 (1958): 21–23; 22.

3. Salvador Dalí, *50 Secrets of Magic Craftsmanship*, trans. Haakon M. Chevalier (New York: Dover Press, 1992), 36–38.

4. Deirdre Barrett, *The Committee of Sleep* (New York: Crown, 2001); D. Barrett, "Dreams and Creative Problem-Solving," *Annals of the New York Academy of Sciences* 1406 (2017): 64–67.

5. Robert E. Franken, *Human Motivation* (Pacific Grove, CA: Brooks/Cole, 1994), 396.

6. Mihaly Csikszentmihalyi, *Creativity: Flow and the Psychology of Discovery and Invention* (New York: HarperCollins, 1996), 28.

7. Csikszentmihalyi, *Creativity*, 28.

8. Engineering Dreams Workshop, MIT, Cambridge, MA, January 28–29, 2019.

CHAPTER 12: WORKING WITH DREAMS

1. Clara E. Hill, *Working with Dreams in Therapy: Facilitating Exploration, Insight, and Action* (Washington, DC: American Psychological Association, 2003).

2. Clara E. Hill and Patricia Spangler, "Dreams and Psychotherapy," in *The New Science of Dreaming: Volume 2—Content, Recall, and Personality Correlates*, ed. Deirdre Barrett and Patrick McNamara (Westport, CT: Praeger/Greenwood, 2007), 159–86.

3. N. Pesant and A. Zadra, "Working with Dreams in Therapy: What Do We Know and What Should We Do?" *Clinical Psychology Review* 24 (2004): 489–512; C. L. Edwards, P. M. Ruby, J. E. Malinowski, P. D. Ben-

nett, and M. T. Blagrove, "Dreaming and Insight," *Frontiers in Psychology* 4 (2013): 979, https://doi.org/ 10.3389/fpsyg.2013.00979.

4. Montague Ullman, *Appreciating Dreams: A Group Approach* (Thousand Oaks, CA: Sage, 1996).

5. C. L. Edwards, J. E. Malinowski, S. L. McGee, P. D. Bennett, P. M. Ruby, and M. T. Blagrove, "Comparing Personal Insight Gains Due to Consideration of a Recent Dream and Consideration of a Recent Event Using the Ullman and Schredl Dream Group Methods," *Frontiers in Psychology* 6 (2015): 831, https://doi.org/10.3389/fpsyg.2015.00831.

6. R. J. Brown and D. C. Donderi, "Dream Content and Self-Reported Well-being among Recurrent Dreamers, Past-Recurrent Dreamers, and Nonrecurrent Dreamers," *Journal of Personality & Social Psychology* 50 (1986): 612–23.

CHAPTER 13: THINGS THAT GO BUMP IN THE NIGHT

1. M. J. Fosse, R. Fosse, J. A. Hobson, and R. J. Stickgold, "Dreaming and Episodic Memory: A Functional Dissociation?" *Journal of Cognitive Neuroscience* 15 (2003): 1–9.

2. T. A. Mellman, A. Kumar, R. Kulick-Bell, M. Kumar, and B. Nolan, "Nocturnal/Daytime Urine Noradrenergic Measures and Sleep in Combat-Related PTSD," *Biological Psychiatry* 38 (1995): 174–79.

3. M. A. Raskind, D. J. Dobie, E. D. Kanter, E. C. Petrie, C. E. Thompson, and E. R. Peskind, "The Alpha1-adrenergic Antagonist Prazosin Ameliorates Combat Trauma Nightmares in Veterans with Posttraumatic Stress Disorder: A Report of 4 Cases," *Journal of Clinical Psychiatry* 61 (2000): 129–33.

4. Ernest Hartmann, *Boundaries in the Mind: A New Psychology of Personality* (New York: Basic Books, 1991).

5. C. Hublin, J. Kaprio, M. Partinen, and M. Koskenvuo, "Nightmares: Familial Aggregation and Association with Psychiatric Disorders in a Nationwide Twin Cohort," *American Journal of Medical Genetics* 88 (1999): 329–36.

6. B. Krakow and A. Zadra, "Clinical Management of Chronic Nightmares: Imagery Rehearsal Therapy," *Behavioral Sleep Medicine* 4 (2006): 45–70.

7. T. I. Morgenthaler, S. Auerbach, K. R. Casey, D. Kristo, R. Maganti,

K. Ramar, R. Zak, and R. Kartje, "Position Paper for the Treatment of Nightmare Disorder in Adults: An American Academy of Sleep Medicine Position Paper," *Journal of Clinical Sleep Medicine* 14 (2018): 1041–55.

8. A. Germain, B. Krakow, B. Faucher, A. Zadra, T. Nielsen, M. Hollifield, T. D. Warner, and M. Koss, "Increased Mastery Elements Associated with Imagery Rehearsal Treatment for Nightmares in Sexual Assault Survivors with PTSD," *Dreaming* 14 (2004): 195–206.

9. E. Olunu, R. Kimo, E. O. Onigbinde, M. U. Akpanobong, I. E. Enang, M. Osanakpo, I. T. Monday, D. A. Otohinoyi, and A. O. John Fakoya, "Sleep Paralysis: A Medical Condition with a Diverse Cultural Interpretation," *International Journal of Applied Basic Medical Research* 8 (2018): 137–42.

10. R. J. McNally and S. A. Clancy, "Sleep Paralysis, Sexual Abuse, and Space Alien Abduction," *Transcultural Psychiatry* 42 (2005): 113–22.

11. McNally and Clancy, "Sleep Paralysis, Sexual Abuse, and Space Alien Abduction," 116.

12. C. H. Schenck, S. R. Bundlie, A. L. Patterson, and M. W. Mahowald, "Rapid Eye Movement Sleep Behavior Disorder: A Treatable Parasomnia Affecting Older Adults," *Journal of the American Medical Association* 257 (1987): 1786–89.

13. Y. Dauvilliers, C. H. Schenck, R. B. Postuma, A. Iranzo, P. H. Luppi, G. Plazzi, J. Montplaisir, and B. Boeve, "REM Sleep Behaviour Disorder," *Nature Reviews Disease Primers* 4 (2018): 19.

14. R. Broughton, R. Billings, R. Cartwright, D. Doucette, J. Edmeads, M. Edwardh, F. Ervin, B. Orchard, R. Hill, and G. Turrell, "Homicidal Somnambulism: A Case Report," *Sleep* 17 (1994): 253–64.

15. A. Zadra, A. Desautels, D. Petit, and J. Montplaisir, "Somnambulism: Clinical Aspects and Pathophysiological Hypotheses," *Lancet Neurology* 12 (2013): 285–94.

16. M. E. Desjardins, J. Carrier, J. M. Lina, M. Fortin, N. Gosselin, J. Montplaisir, and A. Zadra, "EEG Functional Connectivity Prior to Sleepwalking: Evidence of Interplay between Sleep and Wakefulness," *Sleep* 40 (2017): https://doi.org/10.1093/sleep/zsx024.

17. D. Oudiette, I. Constantinescu, L. Leclair-Visonneau, M. Vidailhet, S. Schwartz, and I. Arnulf, "Evidence for the Re-Enactment of a Recently Learned Behavior during Sleepwalking," *PLoS ONE* 6(3) (2011): e18056, https://doi.org/10.1371/journal.pone.001805.

18. C. H. Schenck and M. W. Mahowald, "A Disorder of Epic Dreaming

with Daytime Fatigue, Usually without Polysomnographic Abnormalities, That Predominantly Affects Women," *Sleep Research* 24 (1995): 137.

CHAPTER 14: CONSCIOUS MIND, SLEEPING BRAIN

1. A. Zadra and R. O. Pihl, "Lucid Dreaming as a Treatment for Recurrent Nightmares," *Psychotherapy and Psychosomatics* 66 (1997): 50–55.
2. Keith M. T. Hearne, "Lucid Dreams: An Electrophysiological and Psychological Study" (PhD diss., University of Liverpool, 1978).
3. Stephen LaBerge, "Lucid Dreaming: An Exploratory Study of Consciousness during Sleep" (PhD diss., Stanford University, 1980).
4. S. LaBerge, W. Greenleaf, and B. Kedzierski, "Physiological Responses to Dreamed Sexual Activity during Lucid REM Sleep," *Psychophysiology* 20 (1983): 454–55.
5. M. Dresler, S. P. Koch, R. Wehrle, V. I. Spoormaker, F. Holsboer, A. Steiger, P. G. Samann, H. Obrig, and M. Czisch, "Dreamed Movement Elicits Activation in the Sensorimotor Cortex," *Current Biology* 21 (2011): 1833–37.
6. B. Baird, S. A. Mota-Rolim, and M. Dresler, "The Cognitive Neuroscience of Lucid Dreaming," *Neuroscience & Biobehavioral Reviews* 100 (2019): 305–23.
7. B. Baird, A. Castelnovo, O. Gosseries, and G. Tononi, "Frequent Lucid Dreaming Associated with Increased Functional Connectivity between Frontopolar Cortex and Temporoparietal Association Areas," *Scientific Reports* 8 (2018): 17798, https://doi.org/10.1038/s41598-018-36190-w.
8. T. Stumbrys, D. Erlacher, and M. Schredl, "Testing the Involvement of the Prefrontal Cortex in Lucid Dreaming: A tDCS Study," *Consciousness and Cognition* 22 (2013): 1214–22.
9. U. Voss, R. Holzmann, A. Hobson, W. Paulus, J. Koppehele-Gossel, A. Klimke, and M. A. Nitsche, "Induction of Self Awareness in Dreams through Frontal Low Current Stimulation of Gamma Activity," *Nature Neuroscience* 17 (2014): 810–12.
10. S. LaBerge, K. LaMarca, and B. Baird, "Pre-sleep Treatment with Galantamine Stimulates Lucid Dreaming: A Double-Blind, Placebo-Controlled, Crossover Study," *PloS One* 13 (2018): e0201246.
11. S. A. Mota-Rolim, A. Pavlou, G. C. Nascimento, J. Fontenele-Araujo, and S. Ribeiro, "Portable Devices to Induce Lucid Dreams—Are They Reli-

able?" *Frontiers in Neuroscience* 13 (2019): 428, https://doi.org/10.3389/fnins.2019.00428.

12. T. Stumbrys, D. Erlacher, M. Schadlich, and M. Schredl, "Induction of Lucid Dreams: A Systematic Review of Evidence," *Consciousness and Cognition* 21 (2012): 1456–75.

13. P. Tholey, "Consciousness and Abilities of Dream Characters Observed during Lucid Dreaming," *Perceptual and Motor Skills* 68 (1989): 567–78.

14. T. Stumbrys, D. Erlacher, and S. Schmidt, "Lucid Dream Mathematics: An Explorative Online Study of Arithmetic Abilities of Dream Characters," *International Journal of Dream Research* 4 (2011): 35–40.

15. T. Stumbrys, "Lucid Nightmares: A Survey of Their Frequency, Features, and Factors in Lucid Dreamers," *Dreaming* 28 (2018): 193–204.

CHAPTER 15: TELEPATHIC AND PRECOGNITIVE DREAMS

1. Edmund Gurney, Frederic W. H. Myers, and Frank Podmore, *Phantasms of the Living*, 2 vols. (London: Trübner, 1886).

2. S. Freud, "Dreams and Telepathy," *International Journal of Psycho-analysis* 3 (1922): 283–305.

3. Freud, "Dreams and Telepathy," 283.

4. Gerhard Adler and Aniela Jaffé, eds., *C. G. Jung Letters, Vol. I* (Princeton, NJ: Princeton University Press, 1992).

5. C. G. Jung, "Practice of Psychotherapy," in *Collected Works of C. G. Jung, Vol. 16*, ed. Gerhard Adler and R.F.C. Hull (Princeton, NJ: Princeton University Press, 1982), 503.

6. S. Freud, "Additional Notes Upon Dream-Interpretation. (C) The Occult Significance of Dreams," *International Journal of Psycho-Analysis* 24 (1943): 73–75.

7. Freud, "Additional Notes Upon Dream-Interpretation," 74.

8. Freud, "Additional Notes Upon Dream-Interpretation," 75.

9. M. Ullman, "An Experimental Approach to Dreams and Telepathy. Methodology and Preliminary Findings," *Archives of General Psychiatry* 14 (1966): 605–13.

10. S. Krippner, C. Honorton, and M. Ullman, "An Experiment in Dream Telepathy with 'The Grateful Dead,'" *Journal of the American Society of Psychosomatic Dentistry and Medicine* 20 (1973): 9–17.

11. Krippner et al., "An Experiment in Dream Telepathy with 'The Grateful Dead,'" 14.

12. L. Storm, S. J. Sherwood, C. A. Roe, P. E. Tressoldi, A. J. Rock, and L. Di Risio, "On the Correspondence between Dream Content and Target Material under Laboratory Conditions: A Meta-analysis of Dream-ESP Studies, 1966–2016," *International Journal of Dream Research* 10 (2017): 120–40.

13. C. Smith, "Can Healthy, Young Adults Uncover Personal Details of Unknown Target Individuals in Their Dreams?" *Explore* 9 (2013): 17–25.

14. E. Cardena, "The Experimental Evidence for Parapsychological Phenomena: A Review," *American Psychologist* 73 (2018): 663–77; 663.

15. A. S. Reber and J. E. Alcock, "Searching for the Impossible: Parapsychology's Elusive Quest," *American Psychologist* (2019): Advance online publication, https://dx.doi.org/10.1037/amp0000486.

16. Richard Panek, *The Trouble with Gravity: Solving the Mystery beneath Our Feet* (Boston: Houghton Mifflin Harcourt, 2019); Richard Panek, "Everything You Thought You Knew about Gravity Is Wrong," Outlook, *Washington Post*, August 2, 2019, https://www.washingtonpost.com/outlook/everything-you-thought-you-knew-about-gravity-is-wrong/2019/08/01/627f3696-a723-11e9-a3a6-ab670962db05_story.html.

EPILOGUE: WHAT WE KNOW, WHAT WE DON'T KNOW

1. T. A. Nielsen and A. Germain, "Publication Patterns in Dream Research: Trends in the Medical and Psychological Literatures," *Dreaming* 8 (1998): 47–58.

2. A. Zadra, T. A. Nielsen, A. Germain, G. Lavigne, and D. C. Donderi, "The Nature and Prevalence of Pain in Dreams," *Pain Research and Management* 3 (1998): 155–61.

3. A. F. Commar, J. M. Cressy, and M. Letch, "Sleep Dreams of Sex among Traumatic Paraplegics and Quadriplegics," *Sexuality and Disability* 6 (1983): 25–29.

INDEX

....

Note: Page numbers in *italics* refer to Illustrations.